D0689985

Close To Home

COLORADO'S URBAN WILDLIFE

Close To Home

COLORADO'S URBAN WILDLIFE

WENDY SHATTIL AND ROBERT ROZINSKI
Photo Editors
FREDERICK R. RINEHART AND ELIZABETH A. WEBB
Editors

ROBERTS RINEHART, INC. PUBLISHERS
in cooperation with
THE DENVER MUSEUM OF NATURAL HISTORY
and
THE COLORADO URBAN WILDLIFE PARTNERSHIP

Published by Roberts Rinehart, Inc. Publishers
Post Office Box 3161, Boulder, Colorado 80303
International Standard Book Number 0-911797-70-X
Library of Congress Catalog Card Number 89-64062
Printed in the United States of America

Designed by Ann W. Douden
Copy edited by Patty Hodgins
Typesetting by Kay F. Herndon
Charts and tables by J. Keith Abernathy
Photography production by Gary D. Hall and Nancy J. Reynolds
Illustrations by Joan K. Klipping
Title page illustration by Glen Whitmore
Logo concept by George Anema

Publication of this book was made possible, in part, by
The Colorado Wildlife Commission
and
Martin Marietta Astronautics Group.

Contents

vii FOREWORD
Jay Hair

ix PREFACE AND ACKNOWLEDGMENTS
Wendy Shattil and Robert Rozinski

xi ABOUT THE AUTHORS

1 INTRODUCTION
Elizabeth A. Webb

13 Chapter One
THE SETTING
Jim Carrier

21 Chapter Two
THE MOODS OF THE CITY
Gene Amole

37 Chapter Three
THE URBAN FOREST
Susan Tweit

45 Chapter Four
DESIGNING WITH NATURE IN MIND:
THE BOULDER CREEK PATH
Claire Martin

63 Chapter Five
ACTIVITIES FOR ASPHALT ADVENTURERS
Janine "Robin" Hernbrode

C O N T E N T S

69 Chapter Six
WILDLIFE BY THE YARD:
XERISCAPING FOR WILDLIFE
Jim Knopf

83 Chapter Seven
WILDLIFE WATCHING FOR INSOMNIACS
Elizabeth A. Webb

101 Chapter Eight
BOULDER'S DEER POPULATION
Charles Southwick, Brian Peck,
Anthony Turrini and Heather Southwick

115 Chapter Nine
PEREGRINE RECOVERY IN
DOWNTOWN DENVER
Jerry Craig

133 Chapter Ten
OF HOPE AND IRONY:
ROCKY MOUNTAIN ARSENAL
Gary Gerhardt

149 Chapter Eleven
THE FUTURE OF URBAN WILDLIFE
IN COLORADO
Jim Hekkers

159 Appendix A
URBAN WILDLIFE PHOTO CLUB

173 Appendix B
CHECKLIST OF COLORADO'S FRONT
RANGE URBAN WILDLIFE

Foreword

Beginning in October of 1988, at a little before 7 p.m. on the first Tuesday of each month the lights would come on at the Denver Museum of Natural History's Ricketson Auditorium to open a unique gathering: the regular meeting of the Urban Wildlife Photo Club. One hundred or so strong at its inception, this group represented the first effort of the Colorado Urban Wildlife Partnership to engage the citizens of the Denver metropolitan area in celebrating the nature that is, as the title of this book expresses it, close to home.

The story of the birth of the Colorado Urban Wildlife Partnership, as well as all of the activities planned for Urban Wildlife Year 1990, starts in London in November of 1987. In England to accept *BBC Wildlife Magazine's* Wildlife Photographer of the Year award on behalf of her partner Bob Rozinski and herself, Wendy Shattil lamented with her fellow photographers that interpreting nature through photography was basically preaching to the converted; that there existed a largely uninformed (and largely urban) citizenry that had grown quite apart from nature. A solution to enlightening this vast group of city dwellers suggested itself at almost the moment the lament was sounded—it came in the form of a popular book for sale in England at the time titled *City Safari.* Consisting of photographs and text focused on the urban wildlife of London, this book served as the catalyst for the effort under way in Colorado today.

Close to Home is a collection of photographs and essays that depicts the unusual but seldom appreciated wildlife of Colorado's urbanized Front Range. Designed to accord with the Denver Museum of Natural History's new exhibit, this book is a product of the dedicated members of the Urban Wildlife Photo Club and eleven local writers. Perhaps fulfilling the original aim of bringing wildlife issues before the public at large, this book—and the partnership that has made it possible—represents a true grassroots effort to remind humans of their bonds to the Earth.

With the deterioration of rain forests and the proliferation of toxic waste dominating the media as the only "environmental news" worthy of the airwaves, it is refreshing to find a positive regional effort that brings the natural world into neighborhoods, schools and even the workplace. One would think that, if the work of the Colorado Urban Wildlife Partnership could be emulated in every urban area of the country, perhaps the major environmental issues of our time would resolve themselves through a kind of mass public enlightenment. At least the work represented in this book gives us hope.

JAY HAIR, President
National Wildlife Federation

WENDY SHATTIL/BOB ROZINSKI

A golden eagle leaps from a favorite nighttime roost, a tall, abandoned chimney at Rocky Mountain Arsenal. Most of the eagles at the arsenal roost in large cottonwood trees. The eagle's long, sharp, heavy, strong talons are evident here.

Preface
AND ACKNOWLEDGMENTS

WENDY SHATTIL AND BOB ROZINSKI

Seeing a wild animal is breathtaking. It rejuvenates on a level we don't fully understand, yet we feel elated and that is enough. To see an eagle in flight is to feel a sense of wonder or at least a perception that perhaps we're not as superior as we think. Despite computer capability and technical advancement, we still have to board a plane to fly. We wouldn't have a clue how to migrate, and who knows where to place the first twig when building a nest? Relating to a wild animal gives a sense of perspective.

A trip to England led us to discover the respect and awareness the British have toward urban wildlife. Concern for the welfare of migrating toads inspired bumper stickers reading "Help a Toad Across the Road." Perhaps we haven't reached that stage yet, but our downtown Denver peregrine releases beginning in 1988 caught people's imaginations when wild meshed with urban, and in *Close To Home* we introduce city dwellers to more of their wildlife neighbors.

With this book in mind, we started the Colorado Urban Wildlife Photo Club in October 1988. It rapidly grew to 200 dues-paying members. Through the club, photographers and wildlife enthusiasts became acquainted with animals, seeing a unique body of photographic work evolve as their slides of foxes, owls, herons and deer in urban settings came alive on screen at monthly meetings. Technically and substantively everyone's skill at seeing and photographing increased, and by the deadline one year after the club originated, over 2000 photos were submitted for book consideration. Of 100 images selected, thirty-four photographers are represented in this book. Most are photo club members.

There are presently only a handful of cities in the world that recognize the quality of our lives reflected in the gift of urban wildlife. Even fewer consciously act to preserve wildlife habitat, as growth and development encroach upon territory that was home to animals long before any human ever set eyes on it. Numerous images in *Close To Home* reflect the adaptable nature of animals, many of which coexist and thrive as long as their basic needs continue to be met. Red fox live in cemeteries. Black-crowned night-herons fish next to the bicycle path along Cherry Creek. A bald eagle at Rocky Mountain Arsenal overlooks the downtown Denver skyline.

Throughout the world Colorado is known for its mountains and wildlife. For those of us who chose this magnificent state to live in, these were factors we most likely considered in making our decision. Colorado wildlife does not exist only in the mountains. The high plains have traditionally been even richer, with vast herds of pronghorn and bison once roaming the short-grass prairie. Eagles, hawks and owls stalked prey along with four-legged predators like coyotes and foxes. They hunted rabbits, prairie dogs and waterfowl. Every single one of these species is represented in *Close To Home* photographs, having continued to coexist with us, at least so far.

It is possible for human expansion and needs of much of our wildlife to be compatible. With awareness and forethought there is room for most of us. Colorado's Front Range has seen substantial development and more is inevitable. We can make an active decision about the quality of change and become stewards of the animals and the land by taking their needs into consideration before irreversibly eliminating our animal neighbors. Then progress will truly mean advancement.

Our gratitude goes to the Urban Wildlife Photo Club; the Colorado Urban Wildlife Partnership; the Colorado Wildlife Commission; the Zoology Department in particular and the entire Denver Museum of Natural History in general. Thanks are warranted for Rick Rinehart and Elizabeth Webb; all the essayists; Jay Hair; Ann Douden; and the extra efforts of club members Cathy Barnes and J. B. Hayes. One final mention is reserved for *BBC Wildlife Magazine,* whose photo competition and awards ceremony in London allowed the opportunity for the discovery of an idea to occur.

About The Authors

Chapter 1:
Jim Carrier is a reporter for
The Denver Post.

Chapter 2:
Gene Amole is a columnist for the
Rocky Mountain News.

Chapter 3:
Susan Tweit is a plant ecologist and
freelance writer.

Chapter 4:
Claire Martin is a reporter for
The Denver Post.

Chapter 5:
Janine "Robin" Hernbrode
is a former education director of the
Denver Audubon Society.

Chapter 6:
Jim Knopf is a landscape architect.

Chapter 7:
Elizabeth A. Webb is the curator of zoology at
the Denver Museum of Natural History.

Chapter 8:
Charles H. Southwick is a professor of
environmental, population and organismic
biology at the University of Colorado, Boulder.

Brian Peck is a ranger with
Boulder Mountain Parks.

Anthony Turrini is an attorney and wildlife
biologist with the National Wildlife Federation.

Heather Southwick is a sociologist
who frequently assists her husband
in biological field work.

Chapter 9:
Jerry Craig is a raptor biologist
for the Colorado Division of Wildlife.

Chapter 10:
Gary Gerhardt is a columnist for the
Rocky Mountain News.

Chapter 11:
Jim Hekkers is the public support program
manager for the central region of the
Colorado Division of Wildlife.

WENDY SHATTIL/BOB ROZINSKI

City Park offers diverse recreational opportunities for Denverites, from cycling to taking the family for a stroll. Canada geese and wildlife everywhere deserve the right of way whenever possible around human traffic.

Introduction

ELIZABETH A. WEBB

Humans are not the sole inhabitants of cities. Many fragments of nature have survived urbanization, and wherever a patch of city green space exists, so does urban wildlife. All around the city, wayside plants and free-living animals are accommodated in such unlikely places as sidewalk cracks, derelict buildings, irrigation ditches, railroad embankments and skyscraper ledges. You can keep the city out of the country, but you can't keep the country out of the city.

Colorado's original landscape was interconnected; then it was fragmented into smaller and smaller pieces by human modifications. A mere 12,000 years ago at the end of the Pleistocene, the Front Range climate and setting were much like they are today—open plains threaded with river valleys. Montane glaciers were receding. The climate was becoming warmer and drier as we entered the current interglacial period. The prairie was thinly populated by humans and packed with free-roaming herds of large herbivorous mammals. Native hunters and gatherers shared the landscape with bison, flat-headed peccaries, camels, elk, mammoths, deer, muskoxen and pronghorn. As notions of private land tenure took hold, those of land stewardship faded with the prairie. Today's prairie landscape is increasingly dominated by people, cities and agriculture; wildlife is subordinate.

Although urbanization of Colorado's Front Range implies alteration and replacement of natural ecosystems, it can be planned with an eye to retaining cottonwood-willow river bottom, cattail marsh and short-grass prairie. Enlightened urbanization harmonizes the human-built cityscape with the Earth-formed landscape. Profitability and preservation can be compatible.

Because of the wide range of altitudes and ecosystems represented along Colorado's Front Range, its cities enjoy large numbers of prairie, canyon, riverine, foothills and montane wildlife. Around 400 species of common fish, amphibians, reptiles, birds and mammals inhabit Front Range urban areas, not to mention the even more numerous invertebrates and plants. Many wildlife species have disappeared from other cities in the United States. The Front Range urban corridor, because of its relatively recent settlement and large remaining undeveloped areas, has been spared some of those losses. Gone are the bison, wolf and black-footed ferret, but on the urban fringe the mountain lion, black bear and bald eagle remain. These species depend on large natural habitat remnants embodied in such disparate entities as Rocky Mountain Arsenal, Pawnee National Grasslands, Rocky Mountain National Park and Roxborough State Park.

The most rapidly expanding habitat on Planet Earth is the city. By the year 2000, nearly half of the Earth's projected population of six billion humans will live in urban areas. In the United States, 90 percent of the 300 million members of the population will live or work in cities by then. Over 80 percent of Colorado's population lives along the Front Range from Pueblo to Greeley. Each of these cities has a unique cultural history, but all have urban wildlife in common. As Colorado's Front Range population continues to grow, there will be increased attention to the importance of urban wildlife. It is incumbent upon city and suburban dwellers, urban planners, wildlife managers, landscape designers and local government officials to preserve existing urban green spaces now. New residential and commercial developments, military installations, abandoned industrial sites, polluted rivers, overgrazed pastures and vacant lots also have great potential for creating new habitat for urban wildlife. By creating livable habitat for humans and wildlife, we will ensure the quality of life to which Coloradans are accustomed. If we are responsible stewards in our own backyards, then the prospect of maintaining global biological diversity comes a little closer to reality. The best way to predict the future is to invent it.

PHOTOGRAPH LEGENDS

PAGE 5 ABOVE

Burrowing owls line up outside their burrow, which originally belonged to prairie dogs. A prairie dog colony can be seen behind the owls. Many tracts of vacant land support this relationship in the Denver area.

PAGE 5 BELOW

A striped skunk and a biker seem a split second away from a close encounter. In fact, the skunk seemed too startled to react defensively, and the biker whizzed right by. In urban settings, humans and animals have enough space to coexist peacefully.

PAGE 6

Commercial and residential development along I-25 and C-470 appears poised to overwhelm the countryside and sweep away these beautiful mule deer. Scrub-oak habitat west and south of Denver is favored by mule deer, but more is invaded by humans each year.

PAGE 7

Two striped skunks make their nightly raid on a dog dish in Golden. The skunk family, six in all, lived peacefully with a family of six raccoons in the area, but had to be ever watchful for their nocturnal enemy, the great horned owl.

PAGE 8 LEFT

A killdeer steps daintily off a curb. Killdeer seem able to withstand human pressure for two reasons: they nest on gravelly, bare ground (even gravel parking lots), which development produces in abundance; they lure cats, dogs, people, and other predators away from their nests by feigning a broken wing.

PAGE 8 RIGHT

What's that at the end of the rainbow—a wet bunny rabbit? This cottontail at the Denver Tech Center dries out after a brief summer shower. The top of Centennial Airport's control tower is visible just right of the rabbit.

PAGE 9 LEFT

A tomato hornworm, larva of a sphinx moth, munches its way relentlessly toward a prized tomato crop in Littleton. This caterpillar was spared and relocated for being such a cooperative subject.

PAGE 9 RIGHT

A bumblebee gathers pollen and nectar from a dahlia in a suburban Denver garden. Luck, as much as skill, is required to take a sharp closeup photo of a busy subject.

PAGE 10

A bullsnake glides across a parking lot at Chatfield State Recreation Area. The photographer prone among parked cars attracted lots of attention. Nonpoisonous bullsnakes are widely distributed in many habitats in Colorado, including the suburbs. They eat mainly rodents, which they kill by suffocating them in their powerful coils.

PAGE 11 ABOVE

On an autumn evening, this leaf-shaped green katydid flew through the open door of the photographer's home on the eighth floor of a Washington Park highrise in Denver. The veins in its wings mimic the veins of leaves. It is shown on the screen door.

PAGE 11 BELOW

Even the slimy, despised garden slug qualifies as urban wildlife. Slugs live in dark, wet spots in the garden, feeding on dead and living vegetation. They relish dahlias. Slugs are land-dwelling snails without shells.

PAGE 12

A Woodhouse's toad took up summer residence in a nearly empty swimming pool near Golden. It spent its time looking for food, sunning, and splashing around in a rainwater puddle. The owner of the pool removed the toad before winter.

AL WALLS

WENDY SHATTIL/BOB ROZINSKI

5

Close To Home

J. B. HAYES

6

Close To Home

RICK MORRIS/WILDWOOD CREATIONS

7

Close To Home

WENDY SHATTIL/BOB ROZINSKI

WENDY SHATTIL/BOB ROZINSKI

J. B. HAYES

DAVID H. SAKAGUCHI

DAVID H. SAKAGUCHI

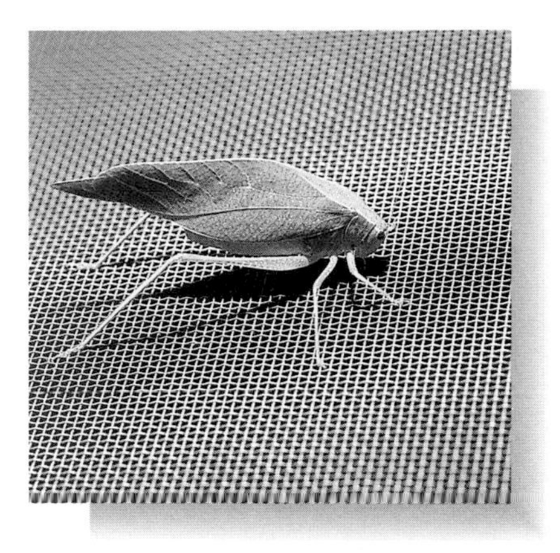

HAROLD ARNOLD

DAVID H. SAKAGUCHI

KENT CHOUN

ONE

The Setting

JIM CARRIER

It rained hard this afternoon with thunder and lightning and sharp gusts of wind that broke off leaves and a small branch from the elm in my front yard.

The tree shouldn't miss the branch. It is in stately shape, big around, tall and spread at the top in the classic arch that made elms such a favorite for shade at the turn of the century. I do not know its age precisely, but I reckon it was planted at about the same time as the silver maple next to it. Their gnarled trunks are roughly the same girth and the house they shade, my house, was built in 1907.

The two trees stand by the street, one on either side of the sidewalk, and spread their shadows and detritus like aging royalty. I don't mind cleaning up after them.

Their arthritic underground knees push the young sidewalk between them, buckling it at the seams, exposing the earth, and creating a catch for debris that falls and blows in. Grass and weeds have invaded those seams: bluegrass from the yard, an errant dandelion, creeping pigweed.

The crack also is a gate to a tiny kingdom that came alive before today's rain stopped. As I watched through a window shade, a robin bounced along the sidewalk's edge, pulling worms that had crawled up for air. A flicker joined it, digging a hole with its beak, flinging wet leaves and tufts of grass and nibbling at ants that had colonized the crack. The flicker shoved its beak deep inside the seam to probe for food. The rain had brought out a feast for these birds, aided by trees planted by humans and a concrete sidewalk that, despite all its intended imperviousness, has been unable to suppress the persistence of nature.

This little tableau of urban wildlife probably was occurring everywhere in the Denver metropolitan area. After 125 years of "civilizing" this landscape, we have failed to run out the wildlife. They have taken root just as we have—in many cases *because* we have.

The flicker, for example, is a woodpecker that lives around trees. It makes no distinction between a natural forest and trees planted by homeowners. The fox squirrel that runs along power lines behind the house must avoid predators as urbane as house cats, but predators the cats are nonetheless. And the magpie that chatters from the alley scavenges there, perhaps finding better pickings than in its normal place in the wild.

The urban setting, from older neighborhoods like mine to the suburbs that creep across the prairie and foothills of the Front Range, eradicates much native habitat. But it also creates pockets for wildlife, sometimes by design, often without intention.

The habitat varies just as neighborhoods do. From what I know of the natural world, that means that the wildlife varies, too. If my sidewalk could be a feeder for birds, imagine what a city park must hold—or a boulevard bordered by flowers, a restaurant's dumpster, the Highline Canal, Cherry Creek Reservoir. In an expanse as wide as the Denver metropolitan area, the possibilities are exciting.

I opened the door to my front yard, quietly stepped around the busy birds and set out to see what the city had to offer.

....

The old trees still dripped and the sidewalk carried a rivulet of water toward the corner of my property. It ran past a neighbor's wall. On the street side manicured bluegrass lay, uniform and lifeless, like a mat around the lot. Inside the wall a lilac bush spread like a tree, hard against a spruce and two aspen. I could hear birds deep in the bush.

The next yard contained a catalpa, looking equatorial with its broad leaves and snaky seedpods. Across the street two young oaks grew vigorously. And where the water ran off the sidewalk and disappeared into the corner storm drain two old cedars stood like sentinels.

I walked past a red maple, a Russian thistle, a plum tree, a blue spruce, a locust—an exotic forest. The wind that followed the storm shook the rain from their crowns with a wet whoosh. I could have been in a woodland anywhere. When I stepped onto Speer Boulevard, the noise of cars startled me.

Two men in black Lycra whizzed past, their heads down, pedaling bikes hard. They disappeared down a ramp. I followed and found Cherry Creek running between concrete banks. It was noticeably cooler, perhaps by 10 degrees, along the creek. The air was moist, and beneath the viaduct nearly dank.

Water from the storm drain gushed out of a pipe in the concrete and fell onto the rocks below. A small pool swirled with leaves and worms washed down from my street. In the murk, two small goldfish darted at my reflection. In the mud around the rocks there were dog tracks, sneaker tracks, and the tiny sharp paw marks of a fox squirrel. Overhead, swallows darted through the shadows, chasing airborne bugs. One lit on a mud nest stuck to a concrete beam and the screams of three chicks erupted, their black-and-beige bodies nearly overflowing their gray house. The mud had been scooped from the creek bottom beneath the viaduct and its humming traffic.

I had been drawn to this waterway just as the wildlife had been. The stream ran wedged between the walls of a flood barricade, and, for the most part, below the street, out of sight and mind of city folks. Yet it still provided the essentials for life: food, water and shelter.

But barely. Where I stood, the creek bed was shorn of nearly all shrubbery. The grass had been cut along the bike path. The water in the creek ran shallow across a smooth sand bottom. Except for some minnows stirring in a tiny eddy and skitterbugs rowing on the surface tension, the creek looked pretty sterile.

Just upstream, though, where a fence blocked my path I could see the banks thicken with life. A big aspen tree had fallen in disarray. A bird nest sat near the top of another aspen. I slipped around the fence, into a thicket of willows. Their roots were exposed along the banks, holding the soil. This was fox, skunk and raccoon country. I pushed through the saplings along what looked like a game trail running parallel to the creek.

Suddenly the thicket ended at a lawn. The grass was sheared like a crewcut. Funny-looking blobs lay on the grass where long slices were cut in the sod. I had found the fifth tee of the Denver Country Club.

The club, 200 acres of luxurious open space in the middle of the city, also marked a change in habitat on Cherry Creek. Downstream, the creek lay between flood barricades as it ran through the city. Its banks were clipped regularly. There was little life there. Upstream, the creek went wild. Willows clogged the banks; shady cottonwoods loomed over the water. There was less erosion, and much more wildlife.

Muskrats live on the country club grounds. So do rabbits and owls. In the 1970s, a den of red foxes got some local publicity. Within a week, someone had shot them, to the extreme consternation of club members. Occasionally mule deer and elk drift in from upstream, according to Mike Fiddelke, the club's general manager. When animals show up on the fairway, they are allowed to play through.

In the midst of the fastidious grooming, Cherry Creek's flora is allowed to grow natural. The weeds and willows are thick as dog hair. The cottonwoods are huge. Wild asparagus and prickly lettuce spring up, and Rocky Mountain bee plants hold water from rainstorms in tiny cups formed by the leaves and stem. It is heaven for wildlife. The tall grasses provide shelter; the weeds produce many seeds. An intricate food chain thrives in the thicket: seeds, butterfly larvae, grasshoppers, rodents.

Beaver so far have stayed off country club property, but their gnawings are visible just up Cherry Creek, across University Boulevard, and in the shadow of a new shopping center. Freed of its concrete banks, the creek here looks like the country.

As I poked around in it, a downy woodpecker flew over the head of a construction worker eating lunch and disappeared into a hole in an old cottonwood. Both beings were here for the same reason, I suspect. Both found in the streambed a bit of respite.

In an unlikely place for wildlife, creatures had found a niche. In an unlikely place for a wildlife lesson, I had found one. The greater the neglect, it seemed, the greater the habitat. A vacant lot gone to weeds could support more life than a perfect lawn. And open space gave wildlife room to roam, to hunt, to hide, to nest. The duffers on the back nine were just another hazard.

Cherry Creek had become for me the artery to the heart of urban wildlife. Following it, I found the heart of the city, too.

....

Denver was founded along Cherry Creek at the spot where it flows into the South Platte River. In 1860 the confluence was a flat watering hole surrounded by short-grass prairie.

Cherry Creek, flowing from the southeast, drained a land that, with only 14 inches of rain a year, baked in the summer. Buffalo grazed, pronghorn gamboled, and both drank at the river's edge. The water spilled over the banks when thunderstorms from the mountains drenched the grasslands. The South Platte River, running northeast, drained the creeks of the Front Range, flooding in spring and later flowing sluggishly, the consistency of molasses.

Outside the watercourses, the confluence was a dry old place in the rain shadow of the mountains, which stripped most of the moisture from the westerly weather systems. The earliest drawings of Denver show a landscape dominated by the two streams with a few cottonwoods along the banks. A lone cabin began the settlement. In one drawing an owl sits beside a prairie dog town. A deer head is mounted over the cabin's door.

Denver grew as a supply and trade community for both the mining in the mountains and agriculture on the plains. In a broad sense, the city is ideally located for wildlife, on the "edge" between the geographic features. Within 30 miles of Denver, the land rises 8,000 feet, creating moist forests, meadows and alpine tundra. The uplift is cut by deep ravines that flow with snowmelt most of the year. Such variety creates a rich array of wildlife, almost 400 species of fish, amphibians and reptiles, birds and mammals.

As Denver grew in the 1800s, the order that only humans bring began to invade the landscape. Streets were laid out in a grid. Bridges leaped Cherry Creek. Homes began their march along the streets. And in front of them, in picketlike rows, new trees were planted. It was the beginning of the urban forest. By the turn of the century, when the trees along my street were planted, the confluence of Cherry Creek and the Platte River could no longer be seen in city pictures. It had been hidden among the rising structures of the Queen City of the Plains. More visible were the trees rising above the homes.

But the water was not forgotten. The city fathers had learned to bend it to their will. As early as 1867 City Ditch was flowing away from the South Platte to water yards, gardens and trees. Where the ditch ran, according to the contours of the ground, Denver's first parks were built: Washington and City Parks as water parks, with large lakes at the center. As the city spread in all directions, so did the waterways. Reservoirs were created along the many streams that flowed into the Platte to save the waters of the spring floods for the rest of the year.

Every suburb that surrounded Denver built reservoirs and ditches. Every mountain community that began creeping up the foothills built elaborate water and flood control systems.

An incredible array of waterways was created. Largely lost in the grid of buildings and streets, and missing from most modern maps of the city, these waterways still serve the original purpose: the creation of an oasis on a semi-arid plain.

The new oasis was not lost on wildlife. The urban forest became home to dozens of bird species not found on the short-grass prairie. Mammals normally confined to the edges of the mountain forests ventured into the suburbs. Nonnative fish and reptiles found their niches, too. Eastern birds like the blue jay moved westward along the rivers and found homes in Denver. The fox squirrel moved in. Houses weren't complete without house mice. And raccoons, which rarely existed in the foothills 50 years ago, began exploring the alleys.

As the city grew, a number of species disappeared, unable to cope with civilization. Buffalo was one. Pronghorn was another, relegated to remnant prairie to the east. That process continues today. A pronghorn herd grazes where a new airport will be built, and its members will no doubt be victims of "progress." Pronghorn are what biologists consider "specialists," limited to a narrow lifestyle.

The urban species tended to be "generalists,"

capable of adapting to or coping with a variety of conditions and habitat. The coyote is a prime example of a species capable of coexisting with civilization.

The hundreds of species of wildlife found in the Denver area today all have one thing in common: the need for food, water and shelter. Where that exists, they have established themselves. The Colorado Division of Wildlife estimates that 85 percent of the wildlife in the urban landscape lives along the waterways.

Over the course of a summer I made several random searches for wildlife in the metropolitan area, and I was constantly surprised by the variety and vigor of the creatures. Despite our best efforts to pave the Front Range, they have found places to live.

•At the city's birthplace, Confluence Park, a waddle of young mallards paddled up Cherry Creek, feeding on submerged plants. Swallows flicked the air along Speer Boulevard and under the Fifteenth Avenue bridge. A doublecrested cormorant flew up the Platte toward Mile High Stadium. There were carp, channel catfish and sunfish in the water. A flock of pigeons peeled away and headed across a railroad yard in search of their adopted home, the warehouse district. A flock of 18 white pelicans swirled in formation over the confluence and headed north. A jet roared above them; a fire engine wailed. But cicadas buzzed just as loudly in a cottonwood.

•At the Division of Wildlife office, stuck in an industrialized section on north Broadway, red foxes and mule deer could be seen along with almost constant tractor-trailer traffic. Just to the north of the office, Clear Creek flows toward the South Platte beneath Interstate 25. Now, whenever I hurtle by at 65 mph, I watch for the trees along it and imagine the life in its shade.

•At the Denver Tech Center, I saw burrowing owls in prairie dog towns on land that is set aside as open space. Voles fed on lawns in the reflection of shiny glass buildings. Hawks wheeled against the skyline. In a white petunia flower bed near the entrance to an office building, a cottontail nibbled at something. Butterflies were everywhere. A shrike sat on barbed wire fence that surrounded the reservoir. To the credit of its developers, the Tech Center has dedicated open space that wildlife has adopted. But there is a big difference between the mowed "open space" with scattered trees seen around many buildings and parks and the unkempt, rather abandoned areas of fields and hedgerows that wildlife need.

•Within yards of the Aurora Mall, prairie dogs are active on land owned by Prudential Insurance Company. On the outskirts of Aurora, where townhomes end and the prairie resumes, sloughs full of cattails, bullrushes and sedges have been allowed to grow between housing developments in foresighted planning. In them, I flushed a red-winged blackbird, peeping frogs by the score and a vesper sparrow.

•In Littleton, a young girl startled me by picking crawdads from an artificially created waterway that channeled runoff from the blacktop streets. She fished among the rocks placed there by a contractor, on a waterfall of concrete and netting and PVC pipe, and pulled from the waterway specimens of one of nature's scavengers.

•A steady flow of news stories from the Front Range reports on urban animals so common that some of them have become pests: beaver on Bear Creek, jackrabbits at Stapleton Airport, deer in Boulder's backyards. There are foxes in Crown Hill Cemetery, night-herons in City Park, pheasants at Chatfield Reservoir, bald eagles at Barr Lake. If further proof is needed, consider this: fishing, hunting and trapping still are allowed in places on the urbanized Front Range.

....

The concrete jungle is not much for supporting wildlife. It's not a very pleasant place for people either. Malls, for example, do not support much wildlife, except for the occasional house sparrow picking at French fries dropped in the parking lot.

But within and around that jungle there are soft spots and wet spots—habitat. In Denver, humans have taken a confluence surrounded by a desert and made it shadier in summer, warmer in winter, and considerably wetter. It has become a comfortable home for millions of people.

One can decry the growth and development of a spreading Front Range, the loss of prime prairie, riparian and foothills habitat, the draining of mountain waters to slake our thirst. The examples of our blundering, ignorant ways are enormous.

Protest is not the purpose of this book. Rather, it is to celebrate nature's ability to compensate for our blunders, and to applaud efforts by some people to take our kindred creatures into account. But most of all it is to notice, and to stand in awe at, the bird in the stoplight, the fox in the cloverleaf, the beaver—bless its nuisance hide—on the golf course. These wild animals remind us every day that we are part of the natural world. They connect us to the earth. In doing so, they civilize the city.

WENDY SHATTIL/BOB ROZINSKI

A mule deer doe and fawn, bedded in dried thistles, warm to the brief December sunshine at Rocky Mountain Arsenal.

T W O

The Moods OF THE CITY

GENE AMOLE

Awareness.

Spring had better be on time this year. Humans aren't the only impatient ones, though. A few anemones, tulips and perennial phlox can't wait. How pleasant to savor the first elusive changes of the seasons.

A crunch of grimy snow from the Thanksgiving Day blizzard remains in a shaded area under one of my blue spruce trees. I'm betting it will still be there the first day of spring.

But that's OK. Blue jays are again chattering in my old cottonwood, and I have already seen swarms of robins. Isn't it exciting to spot the first robin?

That reminds me of a newspaper photographer I knew. He will remain nameless to protect the guilty. My pal always enjoyed the coming of spring and taking pictures of the first robin. And do you know something? He photographed the same robin every year.

Not possible? Yes it was. I know it was the same bird because it was a stuffed robin he carried in the trunk of his car with his cache of other photo props. Late in winter each year, he would take the robin from his car, dust it off and perch it on a snowy tree limb or on a rock in the snow.

I can close my eyes and still see those pictures. That perky little stuffed robin had its head cocked to one side as though listening for a worm wriggling somewhere under the snow.

If you squint, you will notice a pale green cast to some of the lawns. Out east in the farm country, the stink of fertilizer is heavy in the air, and the sound of farm machinery echoes across empty fields.

"The prairie melts into the throats of larks," as poet Thomas Hornsby Ferril so eloquently wrote, "and green, like water, green flows into the pinto patches of the snow."

Lordy, I wish I had written that.

But for those of us who are not poets, there are relaxed ambles through Bear Creek Park. Small animals, ducks, terns and other water birds are trying to make sense of new sandbars sculpted by the heavy runoff last year. Beavers have come downstream from the mountains and are engineering little dams in the willows.

Have you noticed the nice edge to the morning air? At dawn, the horizon is so close that its texture seems just beyond touch. And even though the great peaks are still snowy, there are unmistakable signs of awakening. Twigs on tree branches bend easily. The soil is black and moist under the mulch.

There may be a specific time when spring arrives. But for me, it is a transitory experience. Every fragile moment is worth cherishing, each subtly different from the one before. The pleasure comes from simply being aware of their existence.

March 13, 1984

The edge.

It must have been about 10:30 p.m. when I walked back alone from the Hilton Hotel to the parking lot behind the *Rocky Mountain News.* Only a few cars remained in the lot. My little Volkswagen was somewhere near the center. I was just fishing my keys from my pocket when I spotted something that looked like a little ball of fur.

In the dim light it looked for all the world like a cat. It was sort of hunched up as though it were trying to keep warm in the cold night.

I walked closer, thinking maybe it had been hurt. But just as I was reaching down to touch it, it hopped a few feet away. Yes, hopped away. It was a rabbit. Lordy!

Did you ever see a rabbit downtown? I did. There it was, hunched up again, looking at me with those big brown eyes. That was when I started rubbing my eyes. I looked again. The bunny was still there.

I know what you are thinking and you are wrong. I was stone cold sober. I have seen rabbits before and that was clearly a rabbit—long ears, puff of a tail, the whole shot.

I looked around to see if anyone else was near. Not a soul. Just the old geezer and the rabbit. That's when I began to feel I wasn't alone anymore. I have always thought I was able to sense the presence of those who have departed this level of existence.

I closed my eyes again and saw Lee Casey, columnist and associate editor of the *Rocky* for many years. After his death Jan. 29, 1951, he was cremated. His ashes were placed in the wall of the *News* building, right across the street from where I stood.

Then I felt Mary Coyle Chase was near. She was a former *News* reporter who won the Pulitzer Prize for her play "Harvey," the story of tipsy Elwood P. Dowd and his best friend, a wise and invisible rabbit 6 feet 1½ inches tall. She died Oct. 20, 1981.

I knew and loved them both. Mary always denied it, but many still believe Elwood P. Dowd was in part modeled after Casey. Having seen "Harvey" on both stage and screen, I certainly saw similarities.

The parking lot wasn't empty anymore. It was getting crowded. The rabbit and I were in the company of old friends. The night wasn't cold by then. It was warmed by memories and by almost-forgotten feelings.

The rabbit turned and hopped away toward Colfax and Copperfield's, where the old White Mule bar used to be. I don't know whether I'll ever see him again, but if I do, I'm going to call him Harvey. And if he starts to get bigger and begins to talk...

Am I getting too close to the edge?

March 18, 1984

Flicker.

Oreo and I had been hearing spring birds along Bear Creek for about three weeks when we spotted our first *Ardea herodias treganzai.* He was 4 feet tall and made a postcard picture as he stood in cold water waiting for an unlucky fish to swim by.

OK, so I didn't know the Latin name for the great blue heron until I looked it up. These magnificent water birds start arriving in this area from Mexico or the Galapagos Islands in late February or early March. We don't see many along Bear Creek. They are more numerous near Box Elder Creek, Barr Lake and up near Fort Collins along the Cache la Poudre river.

The interesting thing about them is that they often return to their same nests, usually high in cottonwood trees, repairing them before mating. Sometimes sparrows and magpies occupy the lower part of heron nests, condominium-style.

I hope you are not hurrying through spring. Slow down and be aware of all the wonderful changes taking place in our world. I love it when the first grape crocuses and tulips punch through the dirt, and the silver maple in our front yard starts to bud.

You'll have to excuse me now. Oreo and I are going to slip down to the park to see if we can spot a red-shafted flicker, or maybe even a meadowlark.

April 2, 1989

Tim-ber!

I don't know who counts the local beaver population these days, but it is safe to say the recent estimate of 94 made by wildlife officials is low. I have been walking along Bear Creek in southwest Denver daily for 20 years, and, from what I have seen, I would estimate the number to be more than double that.

Damage to landscaping in the area is heavy. Some 500 trees have been felled along the Platte River Greenway during the past six months. The replacement cost is estimated at $100,000.

We are not just talking saplings. Mother Nature's little engineers think big. In November, they took out a tree that was at least 1 foot in diameter near the Raleigh Street foot bridge in Bear Creek Park. The cottonwood crashed down during the night, falling across Bear Creek, paralleling the foot bridge.

Trying to find homes for beavers hasn't been successful, and I doubt that plans to relocate them this summer will work, either. The beavers apparently love city life and intend to stay. They have worked their way into the center of Denver along the South Platte River.

So-called bank beavers that populate most of Denver's waterways are very difficult to trap alive. They build their dams and store their food along the riverbanks. Rural beavers, found along farm irrigation ditches, are much easier to capture.

Wildlife preservation organizations vigorously oppose kill-trapping the beavers, contending that man created the problem, not the industrious rodents. I suppose that's partly true, since man has landscaped the Platte River Greenway, thus giving beavers new sources of food and shelter.

Inadvertently, man also provides habitat in Denver for raccoons, foxes, muskrats, skunks, cottontails, pheasants, coyotes, prairie dogs and deer. Yes, even deer. My wife, Trish, encountered a big buck this winter along the Bear Creek bike path between Wadsworth and Estes.

If you would have told me 15 years ago that I would have to brake my car in parks to keep from running over Canada geese, I would have said you were crazy. Estimates of the Canada geese population along the interurban Front Range run from 25,000 to 75,000.

Maybe all this inmigration of wildlife means the critters are coming back to reclaim what was theirs. If that's true, it will be a case of the survival of the cutest. Public sympathy is more easily aroused if the animal appears cuddly like a character in a Walt Disney cartoon.

But when rattlesnakes make their annual spring encroachment of suburban Aurora, you can bet that animal-rights activists won't be circulating spare-the-serpents petitions.

February 15, 1987

Canada geese.

I thought of them the other evening when a flock of locals flew right over Depew Street. They couldn't have been higher than 100 feet from the ground. They were gone almost as quickly as they came. I am always excited at the sight and sound of those big birds.

The geese I saw were probably on their way over to Marston Lake. The *Rocky Mountain News'* outdoor editor, Bill Logan, told me it will be late next month before we see big flocks of migratory Canada geese.

Logan said there are about 5,000 resident geese scattered between Denver and Fort Collins. He calls them "common" geese. The flocks get larger each year. Some of the migratories like it so well here they just decide to stay.

I get a kick out of the geese that waddle around Sloans Lake. They seem to thrive on urban living. I have noticed they sometimes get in the way of bikers and joggers. They just refuse to be hurried out of the way.

There was a time some years ago when I didn't have those friendly feelings about geese. It was at that time in life when autumn had a different meaning for me. Autumn was hunting time.

Henry Swan, a Denver surgeon, used to take me out to his lease on the South Platte. He had somehow managed to tow an old refrigerator railroad car out there in the middle of a stubble field. He had fixed up the inside with a small kitchen and sleeping space for a half-dozen people.

We were on one of our annual excursions when I came face to face with three of the biggest, most beautiful geese I had ever seen. But I am getting ahead of my story.

Henry and I drove out to the river on a Saturday afternoon. It was the last day of the pheasant hunting season. As we were driving through one of the countless farm gates, Henry spotted a big rooster pheasant.

We jumped out of the car. His dog started working along the tumbleweed-filled ditch. Up popped the bird. Henry nailed him with one shot.

We had pheasant for dinner that night. Henry, a superb cook, made a white wine and mushroom sauce. We ate the bird, laughed, told a lot of lies and sacked out early.

The next thing I remember was Henry hollering that it was 4 a.m. The temperature had dropped to 14 below zero. Everything was frozen solid, including my feet.

Henry and I stumbled out of that old refrigerator car and into one of the coldest mornings I could remember. The full moon was bright. It glistened on the frosted corn shocks.

We made our way toward the blinds on foot. We must have walked several miles. It was so cold. The dog didn't want to come. It was that cold.

When we finally got there, we noticed that a part of the river hadn't frozen over. We found some cover and settled in, figuring that when it became light, a few ducks might come in.

Just as the eastern sky began to glow, we started to load our shotguns. We had spotted some geese moving along the horizon and decided to use No. 2 shot.

I snapped three shells in my old J. C. Higgins pump. Henry's more expensive Browning automatic was frozen solid. The geese began to move closer. Henry tried to melt the ice on his gun with a Zippo lighter. No go.

All of a sudden, three big and beautiful Canadas began to fly up the river, right toward us. Henry banged his fist against the Browning. Still no go. "All right, Geno, it's up to you," Henry whispered. "I'll tell you when to shoot. Not yet. OK, now lead them a little more. Not yet."

The geese were then so close we could almost touch them. "Get ready. Now, Geno," Henry said urgently. "Now! Pull the trigger, you sonofabitch!" I couldn't do it. I was mesmerized by those magnificent birds. There was no way I could shoot them. They were too beautiful. Too alive.

Henry forgave me. I think.

October 9, 1980

PHOTOGRAPH LEGENDS

PAGE 29

Two buck mule deer greet the sunrise at Rocky Mountain Arsenal. Their antlers are covered with summer velvet, soft, fuzzy skin that protects and nourishes the growing bone. The rack on the deer to the right is unusually flattened and outspread; the other deer's rack is more typical.

PAGE 30

A great horned owl snoozes on a silo. It kept its eyes tightly closed while the photographer came as close as 25 feet. It opened its eyes only when the photographer retreated, and it never did fly. The keen observer can find great horned owls sleeping in trees and on buildings during the day throughout the Denver area.

PAGE 31 ABOVE

Water is as necessary as food in attracting wildlife. Here a raccoon washes its paws before drinking from a water dish provided by a Boulder gardener. The rock keeps this rascal from tipping over the dish.

PAGE 31 BELOW

A mule deer fawn, still with baby-blue eyes, nuzzles up to seedlings in a Boulder garden. It crawled under a fence while its mother browsed nearby. Deer dine freely on Boulder landscaping, annoying some gardeners and homeowners.

PAGE 32

A Swainson's hawk migrating south in fall has a long way to fly—to Argentina. Could it be thinking about hitching a ride partway? Swainson's hawks live in Colorado in summer. Their niche is sublet in winter to rough-legged hawks, which come from their summer homes in the Arctic.

PAGE 33

A grove of large cottonwood trees killed by the rising waters of Chatfield Reservoir is an important rookery for great blue herons and double-crested cormorants. Smaller birds nest here, too, and bald eagles roost here in the winter.

PAGE 34 ABOVE

This common grackle on a newspaper vending machine invites us to anthropomorphize, but in fact it is using the handiest high perch to look for food or danger. Birds on signs had a special appeal to photographers for this book.

PAGE 34 BELOW

European starlings and a pigeon adorn a billboard near West Colfax and Sheridan in Denver.

PAGE 35

A persistent cat spends a lot of time waiting in this tree near Sedalia. The birds have too much sense to use the house; instead, they find other houses that the cat's owners put in less accessible places. Cute as this photo looks, pets can be a threat to wildlife if not properly controlled.

PAGE 36 ABOVE

A band of elk grazes in an Evergreen yard, apparently waiting for the RTD bus to come along. Elk are common around Evergreen and elsewhere along the foothills, and they even venture onto the plains just south of Denver.

PAGE 36 BELOW

These Canada geese waited patiently outside Denver Zoo's until the 10 a.m. opening, then trooped through the gate. With lifetime zoo passes, why should they bother to sneak in by flying?

WENDY SHATTIL/BOB ROZINSKI

Ferruginous hawk perched on railroad car at Rocky Mountain Arsenal.

WENDY SHATTIL/BOB ROZINSKI

DALE PETERSON

WENDY SHATTIL/BOB ROZINSKI

RUTH CAROL CUSHMAN

WENDY SHATTIL/BOB ROZINSKI

GORDON AND CATHY ILLG

WENDY SHATTIL/BOB ROZINSKI

DALE PETERSON

CECILIA T. ARMBRUST

BILL SEARCY

WENDY SHATTIL/BOB ROZINSKI

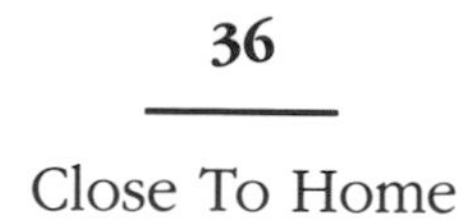

THREE

The Urban Forest

SUSAN TWEIT

I heard a chain-saw engine snarl the other morning and knew without going outside to look that the sound heralded the end for the big old silver maple at the corner. The large blue X spray-painted on its silvery trunk meant that the Boulder city forester's crew had given it the death sentence. Its once-graceful limbs were hollow and broken, the rot-brown innards revealed by gaping holes. It had become a liability rather than an asset.

Silver maples, the shaggy-barked trees whose arching branches cast dappled shade over so many older Front Range neighborhoods, were commonly planted at the turn of the century because of their speedy growth and spreading form. The resultant canopy of tall old trees with outstretched branches is a grand sight, lending a gracious character to the neighborhoods it shelters. Mapleton Hill, for instance, one of Boulder's Victorian-era neighborhoods, is named for its profusion of graceful silver maples.

These old maples, together with thousands of other trees along Colorado's Front Range, form the Denver-Boulder metropolitan area's vast urban forest, the collection of trees and other plants growing along city streets, in yards, in parks and in open spaces. Although "urban forest" at first sounds like an oxymoron, it is not. City and community vegetation may not be "wild," since it is largely the product of planned plantings, but collectively these plants comprise a forest. Like other forest communities, the urban forest plays a key role in its ecosystem: It interacts with and affects local climate, wildlife populations, air quality, water quality and quantity and human behavior.

Looking at urban vegetation in toto, as a part of an ecosystem whose parts and processes are interdependent, is a new idea. Traditionally, city and urban trees have been planted and cared for individually, without consideration of their cumulative effect. But over the past two decades, foresters, municipal officials and private citizens have shifted their perspective and begun managing the forest as well as the trees.

The traditional approach simply could not cope with trends like shrinking wild or open land, increasingly scarce water and other resources, insect and disease epidemics and potential liability for damage caused by urban trees.

Trees shade over half of the ground in Front Range communities. From the air, urban areas in summer look like billowing green islands in a sea of pale ochre grasslands and tilled fields. A closer look shows that the green islands are not homogeneous, but rather a lively mosaic of color and texture, depending on what trees have been planted and how many of each kind there are. While an older neighborhood may have a nearly continuous canopy of one or two kinds of spreading deciduous trees, a newer suburban area may be mostly lawn, dotted with smaller, more sparsely spaced trees of many kinds. Acres of cement and asphalt surfaces often characterize downtowns, but even these monotonous landscapes are interrupted by refreshing green patches of street trees and parks. And where a stream slices through the rectangular street pattern may be a plush ribbonlike canopy of dense riparian (streamside) forest.

The most extensive Front Range urban or community forests are those growing in residential areas. They include street, yard and park trees and other vegetation planted around houses, low-density commercial areas and city parks. The trees of the residential forest are mostly deciduous, many of them brought by turn-of-the-century settlers from their Midwestern and Eastern homes: silver, Norway and red maples; pin, bur and red oaks; American and Siberian elms; horse chestnuts; catalpas; black and honey locusts; green ashes; lindens; pears; crabapples. Others are native to Colorado, either to the plains riparian forests or the mountain coniferous forests: plains cottonwoods and other poplars; willows; Colorado blue spruce and other spruces; Douglas fir; and various pines and junipers.

The form and structure of residential forests depend very much on when the trees were planted. Older forests generally have bigger trees, and therefore more continuous canopies; younger forests are often less dense, partly because tree size correlates to age, but also because smaller kinds of trees have become more popular and increasingly less hospitable conditions in urban areas have made tree growth more difficult. Species planted depend on the fashion of the times: for instance, silver maples, oaks, elms and horse chestnuts were popular around the turn of the century when more patient generations venerated spreading, shady trees; the instant neighborhoods in the sprawling suburbs of recent decades are characterized by more lawns and smaller, fewer trees, including evergreens and fast-growing, low-maintenance locusts, Norway and red maples and green ashes.

Forests of residential areas and parks are home

to a variety of wildlife, the most abundant and commonly seen of which—robins, starlings, house sparrows, cottontail rabbits, raccoons, fox squirrels, deer mice, paper wasps, cabbage white butterflies—comprise a truly urban fauna. These species flourish in urban forests from Mountain Lakes, New Jersey, to Sacramento, California.

Front Range commercial and industrial areas also support urban forests. Composed of street trees and solitary landscaping groupings amid parking lots and buildings, and weedy volunteers in vacant lots and other disturbed areas, these forests are much less dense and more scattered than forests of residential areas and parks. The same kind of trees characterize the forests of commercial and industrial areas, but the composition is very different than in residential forests. Fewer large, spreading species like American elms, oaks, silver maples and horse chestnuts grow in commercial and industrial forests. Planned plantings are characterized by the ornamental cherries, hawthorns, crabapples and others; hardy volunteers like Siberian elms and plains cottonwood colonize disturbed places.

Forests of commercial and industrial areas grow in the most hostile of urban habitats. They manage to thrive despite nutrient-poor, compacted soils; high levels of air pollution; injury from passing cars and trucks and vandalism. Excess heat radiated by surrounding pavement and buildings exacerbates the already arid environment. They compete for light, water and growing space with city infrastructures: utility lines or underground cables; water and sewer pipes, sidewalk, street and parking lot paving; buildings.

Since commercial and industrial area forests are discontinuous and scattered, their wildlife populations are less diverse than those of residential areas. Large numbers of the few species that can survive in the limited niches of city habitats dominate: pigeons, house sparrows, European starlings, fox squirrels and Norway rats.

Not all Front Range urban forests have been planted. Wild forests either surrounded by, or at the fringes of, human development become part of the urban forest. The most diverse of these are the native cottonwood riparian forests. Where perennial streams or large, non-channelized irrigation ditches thread their way through urban areas, riparian forests provide linear wild oases. A vigorous riparian forest includes a dense overstory dominated by plains cottonwood, peach-leaf willow, box elder and green ash, and a diverse understory including wild plum, shrubby willows, wild rose, hawthorn, currant and chokecherry. The varied assemblage of different kinds, sizes and shapes of native plants plus free-flowing water equals some of the most productive wildlife habitat along the Front Range. Riparian forests provide habitat and traveling corridors for such an array of inhabitants as great horned owls, white-tailed deer, tiger salamanders and endangered plains leopard frogs.

The importance of these relatively lush ribbons of native vegetation to urban ecosystems can hardly be overstated. Not only do they provide a touch of wildness in the midst of town, they are crucial to the health of the waterways they line, helping to keep streams a constant temperature, holding stream banks, and slowing, filtering and cleaning urban runoff. They also provide recreation opportunities for thousands of people within walking distance of their offices and homes.

Urban forests are also found where Front Range populated areas finger into native conifer forests. The fluctuating interface between developed and wild lands becomes urban forest when city plantings and wildlife largely replace the native communities. Otherwise, forests of the urban/wild land margin remain wild ecosystems, subject to natural processes—cyclic fire, drought, and outbreaks of bark beetles and other tree predators—rather than urban problems.

Urban forests are valuable community resources. In many ways, some obvious and easy to assess, others less tangible, they improve human-designed and -built habitats.

Contrary to former president Reagan's belief, trees do not pollute—quite the opposite. In the process of making and using their own food, trees release oxygen, without which we could not survive, and take up and fix carbon dioxide, one of our primary industrial pollutants. Carbon dioxide, a natural component of our atmosphere, is one of the gases that create the greenhouse effect, keeping the Earth warm by providing a sort of canopy over the globe that retains solar radiation. The past three centuries of human activities have brought a rapid increase in the amount of carbon dioxide in the atmosphere. Too much carbon dioxide could increase the temperature in the "greenhouse," causing a disastrous long-term global fever. Urban forests can help offset the increase: healthy, fast-growing trees fix from 25 to 50 pounds of carbon dioxide per year. Trees also clean the air in other ways: They trap dust and other particulates on their leaves and release pleasant fragrances, diluting urban odors.

Urban forests are giant sponges. Dense plantings of trees and shrubs can act as sound breaks, absorbing noise pollution the way windbreaks absorb the force of moving air. Further, trees mask urban sounds with the rustling of their leaves, the creaking of their branches, and the noises of birds and other wildlife they shelter.

Forests help stabilize urban water cycles by absorbing excess rainfall. Trees' leafy canopies intercept precipitation, slowing down and trapping water so that it can sink into the soil. Water that passes through the soil on its way to streams, lakes and underground aquifers moves slowly, reducing the chance of flooding and maintaining more constant flows. Water taking this route arrives cleaner. On the way it is scrubbed by the soil, leaving behind many pollutants: fertilizer and pesticides from lawns and gardens; oil and other residues from roads, parking lots and industrial areas; and other by-products of human lives. Tree and other plant roots bind the soil in place and help maintain a healthy, porous soil structure.

Urban forests save energy. Acres of paving and heat-absorbant surfaces make urban areas simmering heat islands in the summer, when urban temperatures average 5 to 10 degrees F hotter than surrounding areas. Enter trees. One deciduous tree shading the south side of a structure can have the cooling effect of five average-size air conditioners running 20 hours per day. Evergreen trees planted to act as windbreaks and to create dead air spaces on the shady sides of structures act as insulation, reducing both summer cooling and winter heating needs. Trees planted to reduce the energy needs of buildings can save 15 times the amount of carbon dioxide that each tree can fix per year. (Less energy used equals less carbon dioxide released by fossil-fuel-burning power plants.)

Urban forests have other benefits. They provide habitat for a wide variety of wildlife. Boulder's deer population comes to town to dine on and take shelter in the succulent, diverse urban vegetation, as do a host of less obvious but equally interesting species. Urban forests increase property values. Mature trees add anywhere from 3 to 20 percent to the value of a property, depending on the age, size and number of trees. And trees are beautiful. They draw people outside to walk or sit in their shade; to admire their diversity of form, color and texture; to appreciate or study the wildlife that they support. They can define a neighborhood, drawing together a community from individual dwellings; link downtown streets with parks; delineate a shopping district. They soften the hard edges of built environments, bringing life to urban and city areas.

But as the demise of the old silver maple at my corner illustrates, urban forests have drawbacks too. Most likely whoever planted the slender maple sapling thought only of how fast it would grow and the cool shade it would soon provide for the yard and porch swing of their stone bungalow. No doubt they did not imagine that over the next 80 years, its swelling roots would shatter a succession of sidewalk slabs, first the original salmon-colored sandstone sidewalk pavers, then replacements of grey concrete. Trees inevitably grow, and then just as inevitably disintegrate with age, facts more often than not ignored at planting time. Growing tree roots can break up sidewalks and driveways, crack building foundations or invade sewer or water mains. Growing limbs can tangle with utility wires or roof overhangs. As trees age and grow weaker, their branches or even whole trunks can fall in wind and snowstorms, crushing whatever they land on—cars, houses, people.

Nor could my silver maple's planters see that their tree, grassy lawn and neat flower beds were part of a wholesale conversion process. Native plant communities, evolved over thousands of years to cooperative associations adapted to the local environment, were replaced over several decades by arbitrary assemblages of plants chosen at human whim. The resultant urban plant communities are strikingly homogeneous, reflecting a national taste for turfgrass lawns, shade trees and shrubs. The homogenization of urban vegetation set the stage for continentwide population explosions of species able to adapt to the newly created urban forest habitat and for the displacement of the native fauna. Witness the explosion of European starlings, introduced to this country in New York City during the late 1800s by a group dedicated to bringing to the North American continent all of the birds mentioned in Shakespeare's works. Starling populations grew slowly until the mid-1900s, sticking to East Coast urban areas. Suddenly, with the spread of suburbs and feedlots across the country, starling populations exploded westward, reaching California in the 1960s. (House sparrows, also ubiquitous in urban areas and also introduced by the Shakespearean avifaunists, have a similar history.)

Replacing the thrifty, drought-adapted native vegetation with a forest of deciduous trees and lawns brought another problem. Trees like maples, locusts, lindens, oaks, elms and others not adapted to the arid climate lose a surprising amount of water vapor in respiration. (Along with vitally important oxygen, plants respire water vapor.) Not only are these newcomers heavy consumers of scarce water supplies, but a forest of them can change the local summer climate. The water vapor given off by a deciduous tree canopy raises the relative humidity of the air, perhaps doubling or tripling the humidity. No surprise, then, that the summer air in well-forested Front Range towns sometimes feels more like that of the sticky Midwest than arid Colorado.

Nor did the planters of my silver maple consider the danger of imposing on their neighborhood a monoculture—an even-aged, single-species forest. Their tree and the other magnificent shade trees of its area, planted within a decade of each other, would also all become decadent in the same span nearly a century later, threatening to denude a neighborhood cherished for its green ambience. Aging is not the only problem in monocultural forests: Insect or disease outbreaks can also be disastrous. Consider American elms, another popular street and yard tree, planted to shade miles of Front Range streets. After Dutch elm disease reached Colorado in 1948, tens of thousands of American elms in Front Range forests died (Denver alone has lost at least 15,000, over half of its American elms), leveling whole neighborhood forests in just a few years.

As they filled in soil around the sapling and tamped the surface down, the silver maple's planters couldn't have imagined that nearly a century later, after the street was widened and stop signs installed, their prized tree would be in the way. No longer an asset, its old and rotting body would pose a hazard to passing traffic and have to be removed.

Urban forests are in trouble. For every four trees that die, only one is planted. Denver's urban forest alone has lost close to half of its trees in recent years. Those that are planted face increasingly shorter lives—an average of only 25 years—in an ever more stressful environment characterized by polluted air, competition for scarce space and light, compacted and poor-quality soil, and conditions aggravated by heat-absorbing surfaces, vandalism, poor maintenance, diseases and other ills. While urban forests are shrinking, urban areas are growing. Seven out of ten of us in the United States live in metropolitan areas. The figure is eight of ten in Colorado, most of us concentrated in urban areas along the Front Range. As metropolitan areas grow, people continue to move out of central cities to greener urban/suburban areas, increasing the competition for space. Hundreds of thousands of acres of forest are lost to development each year.

Front Range communities were endowed with splendid forests by city founders, building a legacy of care for urban trees. Turn-of-the-century homeowners and city officials planted thousands of maples, elms and other shade trees, many of which, like those in Denver's Park Hill or Mapleton Hill in Boulder, still cast their graceful shade over whole neighborhoods. (Many of Denver's older trees date from then-Mayor Robert W. Speer's tree giveaway in 1913, featuring elms and silver maples.) Colorado Springs, spurred by founder General Palmer's interest in city trees, became in 1910 one of the first communities west of the Mississippi to establish a city forester position. Today, a new awareness of the importance of forests in making urban ecosystems livable is evinced by the formation of groups like Denver's Urban Forest, working to preserve and extend community forests.

Important local resources now, urban forests will become increasingly critical global resources as the world's population grows. As global deforestation continues apace—each year, tropical forests alone lose an area the size of the state of Tennessee—urban forests represent a larger total of world forest land. They can help stem the increase in atmospheric carbon dioxide: Trees planted in all the hundred million available tree-planting spaces in United States urban areas could reduce carbon dioxide emissions from energy production by about 18 million tons per year and fix some 1.5 million tons. Such a planting would be a small but efficient step toward reducing the global overproduction of 2.6 billion tons of carbon dioxide per year. We may someday depend on urban forests for many more returns than we currently do. Integrated into self-sufficient communities, they could help balance community energy budgets by lowering energy needs and providing for fuel wood, assist in water filtration and treatment systems, meet recreation needs and provide jobs in urban tree farming and forest care.

As far as we know, our Earth nurtures the only life in the galaxy, perhaps the only life in the whole of our vast universe. Surely we, the first species powerful enough to destroy all life, are also wise enough to learn how to perpetuate it. Learning how to create and maintain healthy urban forests is a first step. Urban forests are clearly more than just pretty trees; they are an integral part of the urban environment in which the majority of us live and where our children grow up and learn the values and beliefs that direct their lives. Such forests sustain our physical lives by cleansing our air and water; they nurture our continuing culture by filling our urban spaces with beauty.

Uniquely human habitats, created by humans for humans, urban forests and urban environments are our reflection. As we treat them, so we treat ourselves. Our challenge is to find within ourselves the grace and wisdom necessary to manage them well.

....

The next time I walked by the corner where the hoary old silver maple had stood, it was gone. All that remained were a pile of fresh wood chips and snakelike silver roots flowing out from where the shaggy trunk had risen. Its demise was marked by a blue plume of chainsaw smoke and the too-sweet odor of rotten wood. Next spring, the city forester's crew will plant a sturdier Norway or red maple to grow up and shade the corner. Goodbye, old tree; hello, urban forest!

FOUR

Designing With Nature In Mind: THE BOULDER CREEK PATH

CLAIRE MARTIN

Of the four portholes framing the opaque gray-green water at the Boulder Creek Stream Observatory, the third window, set slightly higher than the others, tends to offer the best view of the drifting trout. They filter like ghosts through the dim, quiet water. To the people who pause to look at them, the fish seem unhurried and impervious in their silent world.

It is strangely moving to study the people staring through the portholes, as if they were watching a quartet of televisions all tuned to the same channel. Most people get bored within a few minutes and leave, but others stay, rapt. They peer into the cloudy water behind the glass long after others have come and gone, as if the wet shadows held the answer to a question they cannot quite frame. Or perhaps they are engrossed because the portholes, set in their concrete frame, offer a steady look at the separate natural world that we often miss just because it has always been there.

Until the Boulder Creek Path opened in 1987, that was also the way many people seemed to feel about the creek. Except in the parks and other manicured pockets where landscaping tamed the tangled underbrush on the creek banks, few people spent their leisure time wandering along the creek.

Among other things, people feared the transients who lingered there. When a pedestrian and bicycle path along Boulder Creek was first being discussed in the early 80s, some Boulder residents whose homes abutted the creek objected that the path would bring uninvited visitors onto their property. As a matter of fact, it did—not the scruffy itinerants that irate homeowners originally envisioned, but well-heeled real estate investors who made offers on a number of creekside homes. Land once worth $60,000 sold for at least twice that after the path was built; new homes there command a minimum of $250,000.

When it opened, the Boulder Creek Path was greeted almost instantly as what *Denver Post* reporter Bruce Finley called "a model of economic, aesthetic and cultural success." City planners from San Francisco to Vermont have praised the Boulder Creek Path and expressed such interest in building similar greenways that Gary Lacy, who designed the path, eventually left his job with the city to work as a consultant for other greenway projects. When, after a 14-state search, US West chose Boulder as the site for its new research facility, a company spokesman told Finley that the greenway was "obviously a factor" in deciding upon Boulder; in fact, the US West facility is being built along the greenway, near the University of Colorado.

What has made the Boulder Creek Path so popular is a combination of its location—much of the path follows the creek as it threads through to one of the few thriving downtown areas in America—and its designers' attempt to be sensitive to the needs of both wildlife and urban life. Boulder has a substantial population of deer, warblers, raccoons and other animals that have followed and congregated near more than half a dozen streams that descend from foothills canyons, a natural staging area for residential and migrating animals and birds.

Even after the town of Boulder began to grow, the animals continued to converge here because of the lush vegetation cover, both natural and planted. Although riparian corridors cover less than 15 percent of Colorado's land area, they host about 85 percent of the wildlife.

So when planners designed the path, they kept these urbanized animals in mind, along with the humans. Jim Zarka, one of the landscape architects who worked on the path, chose certain trees—hawthorn and chokecherry, for example—because they attract wildlife, and planted others, including plum and apple, to "feed transients." That helped console local birders, who were concerned that willows and other traditional habitat for riparian birds were being cleared with an excess of zeal.

Still, some wildlife advocates criticized the path's landscaping as a creation intended more to please human eyes than to provide the habitat and forage that warblers, red-eyed vireos, screech owls, thrushes and other riparian birds require. There was a feeling that the natural creek was being replaced with an artificial one. Creekside brush literally was leveled along the stretch of the path that crosses through the university—a stretch that happens to include much of what Alex Brown, who helps compile the Boulder Audubon Society's wildlife inventory, describes as "what used to be the warbler migration hot spot."

According to previous bird counts, Boulder Creek was one of the favorite layover spots for migrating warblers, as well as a breeding ground for other birds who were seasonal or year-round residents. They were attracted to the dense thickets that hugged the narrow banks and shrouded the creek, creating a place of relative solitude amid the traffic noise.

When that brush was yanked out and replaced by landscaping that allowed a clear view of the creek, both habitat and solitude disappeared, and so did many of the specialized birds.

Brown is among the people with mixed feelings about the Boulder Creek Path. He was sorry to see the warblers and other riparian birds go and believes they may have been unnecessarily exorcised, that the path's architects might have prevented their departure by, say, keeping the path on one side of Boulder Creek instead of having it repeatedly cross the water, disturbing habitat areas on both banks. On the other hand, he wonders whether the increasing strain of urban living eventually might have driven the birds away anyway.

"It's an area under severe urban pressure, and if the path hadn't been built, the human pressure probably would have degraded it," Brown says. "In some ways, building the creek might have actually made the area better than it might have been. One thing that should be borne in mind: In settling Boulder, not all that we've done has been negative. There are many more species of birds here than before, because they were drawn by trees people planted, and reservoirs—open water that wasn't there before. It's hard to say what's good or bad. You can't keep things as they are, but try to make sure each decision you make is the one that does the least harm."

If the Boulder Creek Path isn't nature primeval (and it's not; during the summer, you can look across the creek at the west end of the path and see lawn chairs perched on the banks of private creekside property), even Alex Brown and other critics admit that it's an agreeable place to spend an afternoon. On weekends, the path is crowded with strollers, joggers, skaters and cyclists. Even on weekdays, the path is busy, particularly with cyclists, for whom the path is a kind of interstate through Boulder, an extraordinarily efficient way of navigating downtown and other commercial areas.

Almost any day, you can find people fly fishing (catch-and-release only), and during the summer small anglers gather at the children's fishing pond near Eben G. Fine Park. The pond is the only place on the creek where bait fishing is permitted, although judging from the amount of tangled line dangling from the surrounding trees, the kids catch more branches than rainbow trout.

Farther down the path, the creekbed widens and becomes shallow. On sunny days, you can walk along the creek shallows and watch the waves mold light into shifting shapes that chase each other restlessly across the stones. Just south of the Broadway underpass is evidence that this is a greenway with a sense of humor: The calls of tropical birds float up from subterranean pipes. The sound bewilders passersby who hear it for the first time; they instinctively look up into the canopy of nearby trees and then, incredulously, follow the sound down to the minimanholes. A few feet away another creative project has been undertaken—an enterprising artist acquired the money and permission to enclose sections of the trunks of two trees with transparent boxes. Think of it as a new kind of environmental impact statement.

The path frequently crosses the creek, as it does near the stream observatory, a deep water hole where, as Boulder landscape architect Jim Knopf puts it, "some of the world's grossest trout live." Because of a steady diet of fish food that people buy from the observatory's two vending machines, the rainbows here are undeniably porcine.

Often, when the path crosses the creek, it connects with a tributary bike path. The city is interested in eventually connecting the surrounding tributary trails to the main artery of the Boulder Creek Path. While expanding the tributary trail system might encourage more residents to forsake their cars for their bicycles or walking shoes, environmentalists are concerned that such an expansion would deal another, more telling blow to the wildlife that the Boulder Creek Path forced to relocate.

Would the native vegetation along tributary trails in other riparian areas be torn out and replaced with manicured flower beds and lawns? Would stocked trout and taped bird calls replace vireos and warblers?

So far, nature has been able to accommodate the alterations imposed upon it, but even something as apparently innocuous as a pedestrian path has a marked effect. When willows and other brush were removed to open the view of Boulder Creek, the birds merely moved to other, undisturbed streamside shrubs and trees. But where will they go if those shrubs and trees are leveled? What happens when the old creek is shaped into a new one of our own devising? The larger question is: Is all nature ours to manicure?

Even if it's done with the best intentions, the answer has to be no. There must be respect for the wild things that inhabit this world with us. In the case of expanding tributary trails, it's possible to accomplish something that preserves and acknowledges native habitat while also serving humans' needs. The South Boulder Creek Trail, an unpaved path, is one example of what can be done with thoughtful planning that respects the riparian diversity as well as the human culture.

The South Boulder Creek Trail keeps to one side of the creek instead of repeatedly crossing it. Areas where particularly sensitive birds and mammals live are fenced off, discouraging both cattle and curious humans from disturbing foraging or nesting creatures.

"If we are going to build these paths, then it's better to be sensible and keep people away from the critical places," says Brown, who regards the South Boulder Creek Trail as a model for future paths. "The average person who's walking his dog or riding his bike doesn't really care whether they're in a highly vulnerable place, where wildlife is concerned, or whether they're 50 yards away; they're equally happy. You need to have that thought in mind when you build trails. And usually if people understand why an area is important and needs to be cared for, they'll respect that. Most harm is done by actions that are unthinking. That's one reason that makes Boulder unique, and even the Boulder Creek Path: People do stop to think about their relationship with their environment."

Brown's comments address what has come to be known as "deep ecology"—the effort to pause in our attempts to shape this world into a better place and consider the effect our hands have on the lives they touch. Rather than emphasizing better management of the Earth, deep ecology requires shifting from that traditional, anthropocentric view to a biocentric one in which we are not necessarily at the top of the hierarchy, but walking side by side with what Barry Lopez calls "the culture of bears and the culture of wolves." Deep ecology means adapting ourselves, not the Earth, to a way of living that honors the natural world—valuing life more than lifestyle.

There is already some evidence that such a shift is beginning to occur. The Boulder Creek Path, for example, represented a start that the South Boulder Creek Trail will advance. Here is proof that you can walk side by side with the culture of warblers and the culture of willows and the culture of rainbow trout.

PHOTOGRAPH LEGENDS

PAGE 52

The great blue heron rookery at Chatfield State Recreation Area is bathed in a coppery glow at sunset. In summer, buoys are placed around the rookery to warn boaters away. Tree roots are weakened by ice action each winter, so that the trees will eventually fall down and the rookery will vanish.

PAGE 53 LEFT

An adult bald eagle, told by its white head and tail, perches at Rocky Mountain Arsenal, in sight of downtown Denver. These endangered birds winter at the arsenal and can also be seen around Chatfield and Cherry Creek Reservoirs and Barr Lake. A bald eagle observatory on the arsenal's eastern border is open to the public.

PAGE 53 RIGHT

The moon sets as the sun rises on a regally antlered mule deer at Rocky Mountain Arsenal. In contrast to the sprawl of Thornton, Westminster, and Northglenn in the distance, the arsenal offers relatively undeveloped, uninhabited land with prime deer habitat. Healthy herds of mule deer and white-tailed deer thrive here.

PAGE 54 ABOVE

The curiosity of this endearing young red fox allowed the photographer to approach within 20 feet. The fox inspected her thoroughly, then romped with a sibling and finally fell asleep on the stump.

PAGE 54 BELOW

A coyote gazes upward at a bird on the wing just at sunset. The momentary distraction gave time for this photo of a normally elusive animal. Coyotes are fairly common at the south end of Denver around I-25 and C-470, where this one was spotted.

PAGE 55 ABOVE

A thirteen-lined ground squirrel peers out of a fire pit at a campsite at Chatfield State Recreation Area. His kind earned the name "picket pin" by its habit of standing straight up and motionless when scanning for danger. Its diet includes seeds, grasshoppers, caterpillars, grubs and mice.

PAGE 55 BELOW

A fenced herd of bison munches on hay in the snow at Genesee Mountain Park along I-70 west of Denver. Between 1865 and 1884, Buffalo Bill Cody and many other hunters slaughtered 60 million bison and nearly exterminated the species. Buffalo Bill's grave is not far from this spot.

PAGE 56 ABOVE

A belted kingfisher perches on a bridge pier high above the South Platte River at Hampden Avenue, waiting to dive on an unsuspecting fish. Mt. Evans is on the horizon. Kingfishers live around water and are common near Denver.

PAGE 56 BELOW

A golden eagle hunts from its perch on an aircraft radio beacon at Rocky Mountain Arsenal. In a climate where trees are naturally scarce, all sorts of birds readily perch on manmade objects, as many photos in this book show.

PAGE 57 ABOVE

A buck mule deer finds free room and board in a Boulder backyard. Here he nibbles his host's dogwood bush.

PAGE 57 BELOW

Ring-billed gulls huddle on the ice in Exposition Park, Aurora. Several species of gull live and breed inland. Ring-billed gulls inhabit Denver year-round. A child apparently left its red wagon on the ice, or else vandals threw it there.

PAGE 58 ABOVE

Black-tailed jackrabbits and prairie dogs are abundant at Stapleton International Airport, where they attract predatory hawks and eagles. These jackrabbits gather around anything that will afford protection from feathered death.

PAGE 58 BELOW

A great blue heron in breeding plumage flies past houses along Lee Gulch, Littleton. These shy birds were common summer feeders in this area until construction of a biking and hiking path along the stream drove them away.

PAGE 59 ABOVE

"It's Miller time!" A bumblebee takes a break from gathering pollen near Golden. One wonders if she ever tried the well-known local brand.

PAGE 59 BELOW

A great blue heron strikes a regal pose at Chatfield State Recreation Area, seemingly oblivious to human intrusion. However, these birds cannot be approached closely, no matter how used to people they are.

PAGE 60

A barn swallow feeds its babies flying insects it has caught by swooping low over the ground. Barn swallows are well adapted to living around humans; they build their mud nests under bridges, in barns or, as here, under the eaves of a front porch in Thornton. The barn swallow is the only swallow in Colorado with a long, deeply forked tail.

PAGE 61

This statue in Civic Center Park commemorates westward expansion, to which we owe the presence of pigeons in Denver, among other things. Yet generations of these avian ingrates have heaped contempt on this statue at close range. Be warned, pigeons: you can fly very fast, but peregrines can fly faster.

PAGE 62

Wildlife presence acan be recognized even when the animals can't be seen. Burrows, diggings, tracks, trails, bones, antlers, skins, fur, feathers, shells, droppings, cropped vegetation, eggs and nests, like this one on a statue at Denver's Riverside Cemetery, all tell a story.

SHERM SPOELSTRA

WENDY SHATTIL/BOB ROZINSKI

WENDY SHATTIL/BOB ROZINSKI

ANN B. KNUTSON

GARY A. HAINES

SHERM SPOELSTRA

LYNNE JONES

J. B. HAYES

WENDY SHATTIL/BOB ROZINSKI

BETTY R. SEACREST

AL WALLS

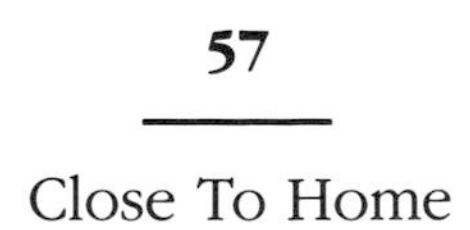

GORDON AND CATHY ILLG

J. B. HAYES

KENT CHOUN

SHERM SPOELSTRA

WELDON LEE

WENDY SHATTIL/BOB ROZINSKI

WENDY SHATTIL/BOB ROZINSKI

KENT CHOUN

FIVE

Activities
FOR ASPHALT ADVENTURERS

JANINE "ROBIN" HERNBRODE

The city of Denver has lovely cascading fountains, period architecture, manicured landscaping and fine outdoor spaces built with people in mind. It also has places that are downright ugly. You might think that wildlife would favor our beautiful parks, choosing their own places as carefully as we choose ours, but even in the homeliest of places we find wildlife.

Last spring, I trained volunteers for the Denver Audubon Society's Urban Education Project. We explored an area south of Colorado's capitol building, along a street with many apartment complexes, a few small eateries, a disconcerting number of homeless people and a sprinkling of paved parking lots. Using as a training classroom a place where beauty is not so obvious impresses on the volunteers that the urban habitat where they will work is both interesting and diverse. The volunteers quickly realize that even the urban schools with only a strip of grass and a couple of bushes have a nearby alley with diverse weeds, a fence line with windblown leaf litter and damp soil where the sidewalk meets the grass. These prove to be hospitable places for wildlife of the smallest kind.

That spring day, we were looking at a strip between two parking lots, probably 10 feet wide and at least a hundred feet long. The steepness of the grade had prompted someone to place some soil-holding rocks there long ago; they anchored a collection of windblown trash and a variety of plants my allergist is fond of calling "Colorado weeds." The volunteers had been given index cards as aids for picking up insects and reclosable plastic bags to put them into. Hunched over this uninspiring hunk of habitat, the group was looking intently for wildlife. We had already attracted the attention of Denver's finest ("Really, Officer, we're just volunteers looking for bugs") when a sudden rustling among the decaying leaves startled us. It was a cottontail rabbit, making for its burrow under a chunk of discarded concrete.

If a rabbit can live in the heart of downtown, your own neighborhood probably has a healthy quantity of wild residents. What follows are some ideas for you and your family or you and your youth group to use to explore the urban landscape as a wildlife habitat and to find out about some of its residents. Most of these activities I have pursued with kids as part of Denver Audubon's Urban Education Project. They were just as exciting to me as to the kids because there aren't any right answers; there are only discoveries to be made.

....

• Have you ever noticed pavement ants? They're tiny little black ants whose mission in life seems to be to build the sandy openings of their nests (one and a half to three inches or so across) in the cracks of sidewalks. Their presence means that the consequence of stepping on a crack not only includes "breaking your mother's back" but also obliterating the lives and days-long labor of hundreds of the little crawlers.

It's fun to try to feed the ants. But what? You might start with cracker crumbs, a tiny blob of jelly and a little peanut butter. Diluting the food with water seems to increase the ants' activity. Just put a small drop of food close to the nest opening and watch what happens. Then try the same food on a hill of larger ants. It doesn't matter what kind of larger ants you find. Just change species and find out the differences or similarities in their tastes and habits. Encourage yourself and your fellow investigators to think ahead and make predictions as to what will happen; then see if it does. Some ants are very aggressive about protecting their own spaces, so be careful that you do not stand in a place where you may be in jeopardy.

• Have you ever seen the creatures living in pond water? Last fall one of the Urban Education Project's groups found a hydra in "Lollipop Lake," a small pond in Garland Park. Unlike the nine-headed Greek mythological serpent with the capital H, Hydra, this being is a freshwater relative of sea anemones, corals and jellyfishes and is no bigger than a pencil dot. In fact, it takes a 39-cent magnifying lens to appreciate much of the life you can dip from a pond's shoreline. A white-bottomed tray or cut-open milk carton with an inch or so of water in it will help you see what you capture. Scrape around mossy rocks, at the base of cattails or other water plants or along a concrete wall with a fine-mesh aquarium net (the kind used for dipping up brine shrimp) and then wash the net in your tray. Examine the wash water with your magnifying lens. Lots of pond water residents are visible without using the lens but you can see their shapes and characteristics better if you use it. Drawing pictures of your finds to take home lets you return the creatures to their own living spaces.

If you want to learn the formal name of your captives, purchase a small pond life guide at the bookstore. The guide will supply names for most of the things you find. For the ones that the guide doesn't cover I prefer to make up a term of my own. If I use words that describe the creature itself, like "green-headed fly" or "shiny rice-grain larvae," I will remember it more easily the next time I see it.

• Have you ever fed birds?[1] Most people feel special warmth for creatures that will take food offered to them. It takes a while for birds to establish a habit of coming to a special place for food. Put some birdseed out in your yard or on your window ledge and wait for the birds to find it—they may need a couple of weeks to learn that your place is a dependable source of meals. Although they will grow accustomed to some distractions, boisterous pets, flapping flags and other disturbances make it harder for them to eat. Let the birds establish a habit of feeding safely before you try to test their reactions by acting on the following ideas.

When you are feeding a steady number of birds, see what happens when you offer two different kinds of food at the same time. If you offer both the usual birdseed and bread crumbs or other alternative food, do they always take one before the other? If you put some of their favorite food in a transparent plastic cup and some in an opaque one, do they find one source faster than the other? If you make a snake shape out of modeling clay and put it close to the food, do the birds react? If you make a loud noise while you are sitting very still, will the birds fly away? Will they flee if you are very quiet but make a sudden movement? I'm sure you can think up more experiments to do, or maybe you would just like to enjoy watching the birds as you feed them. For safety's sake, never try to get any wild creature to take food from your hand.

When you were experimenting with sitting still or making sudden movements and making loud noises or being very quiet, you probably noticed that the birds responded to your behavior. How would you act if you wanted the birds to come close to you and ignore your presence? See how long it takes the birds to accept you when you are acting the way you found most effective. Many naturalists have used this method to study the behavior of wild animals quite successfully.

• Have you thought about grasshoppers as afternoon companions? You could use leafhoppers or crickets if those are the kinds of hoppers most available to you. Discovering the capabilities of different hoppers is a fine way to pass the time and also nets you some understanding of the creatures themselves.

First you must catch a few specimens and put them into a can or box. Cover your container

[1]The concentration of large numbers of birds in the small space around feeders can allow diseases to be passed more easily from one individual to another. Therefore, the Denver Audubon Society recommends limiting feeding to the times birds need it most—late fall, winter and early spring.

with plastic wrap and secure it across the top with a rubber band. Don't worry about punching holes for breathing; there is plenty of air in the can for the hoppers to live for an afternoon. Just keep the can in the shade.

Here are some ideas to get you started on your investigations. Use a piece of string to measure how far one of your hoppers hops. As it hops more, does the distance change either by getting longer or shorter in successive hops? Does it hop the same direction it is facing when you set it down or does it always hop uphill, downhill or at the same level? Research has shown that certain grasshoppers only have enough stored energy to jump a limited number of times before they become unable to continue and must rest. Count how many jumps yours can make. Try hopping your hopper in both the sunlight and the shade to see if that makes a difference in its behavior. If you chase your hopper so it must jump more than once, does it change direction frequently or does it hop mostly in the same direction? When you are finished playing with the hoppers be sure to return them to their habitat. Dead hoppers aren't much fun.

•Have you noticed the activity in a pile of dead leaves? You probably haven't spent a whole lot of time sitting on the curb, but in case the opportunity presents itself on a sunny warm day in winter or spring, you might check out the community of organisms that live in leaf litter. Leaf litter is a seasoned pile of debris that is decomposing before your very eyes. Fence lines, curbs and alleys usually have a good collection.

There is much more activity when the leaf litter is moist, so douse it with a milk jug full of water if the weather has neglected to dampen it for you. If you must wet it down, wait 45 minutes or so for the residents to become active. Many of these creatures are very tiny, and the search for them takes patience and a sharp eye. A reclosable plastic bag to hold your finds for inspection will let you see both sides of a beast without risking a bite or sting. An index card will help you pick some of them up without squashing them. Notice how many different kinds of insects live in leaf litter. Did you rake your yard so thoroughly that you didn't leave a habitat for them at your place?

•Have you noticed how diverse the urban setting is? Wildlife populations follow some of the same rules humans do: The more places there are to live and the more ways there are to make a living in an area, the more the varieties of wildlife that can live there. It's fun to celebrate that diversity by comparing different areas of the city. Pick a nearby city park and compare it with an alley or vacant lot. See how long it takes you to find:

five unusually shaped leaves
five different kinds of rocks
five different seeds
five different insects

five different animal signs (webs, droppings, leaves nibbled by insects, etc.)
five things colored different shades of brown
five things colored different shades of green
five different odors
five different textures

•Have you ever looked at the bottoms of plants rather than the tops? We usually spend so much time concentrating on the leaves of plants that we forget there is variety in the roots too. Find a place where no one will mind if you dig up the plants to see what kind of root systems they have. Vacant lots, alleys, road rights-of-way and other areas that are untended by loving gardeners are usually great places to do this. You can make root research into a game by having your companions cover their eyes while you dig up a couple of different plants. Put a paper bag over the aboveground parts and secure it with a rubber band so the players can see only the roots. Their challenge is to find a matching plant, with only the roots as a clue.

After your mystery plants have been found, you might search for a plant:

with a root system larger than its aboveground stem
with a root system smaller than its aboveground stem
with more branches on the root than on the stem
with more branches on the stem than on the root
with a very shallow root system
with a stem thicker than the main root
with a main root thicker than the stem

I hope you found some examples of the great variety in root types and sizes. Could you predict the kind of root you would find by looking at the top?

•Have you ever fed a spider? Try it at night, using a flashlight. Spiders are common and often tenacious dwellers under the eaves of your house, on exterior walls, inside the garage and throughout your garden. Most of them build webs to catch insects for food and are more active at night than in the daytime. Some webs are so obvious they're easy to find, but many are not so large or thick and require some searching. If you are having trouble, try placing your flashlight at forehead level as you look to help you catch the light reflected in spiders' eyes.

Do not play with black widow spiders, which are poisonous. Black widows have glossy black bodies, usually with a reddish mark on the underside. Spiders inject their prey with a poison to paralyze it and juices to digest it; then they suck the liquid for nourishment. Most of the time the poison and juices cannot do much harm to a human being, but the black widow is a notable exception.

A porch light will attract some insects to feed your spider. Catch some in a reclosable plastic bag. You can use tweezers to take the "food" from the bag and toss it into the web, or you can use your fingers if you like. The trick is getting the prey into the web without touching the web and destroying its delicate structure. It is okay if the insect tangles itself up in the web. Watch what the spider does. You will probably want to try several webs. If you have difficulty getting a response, try a different kind of spider.

You have probably dealt with the stickiness of spider webs before. Once they are on your broom they are not on the wall, but then what do you do? Have you ever thought about why the spider does not stick to its web? Using a thin broomstraw, see if you can tell if all of the threads of a spider web are equally sticky. Be gentle, but try not to feel too guilty if you accidentally destroy a section of web. Spiders are constantly repairing and replacing their webs after natural disasters. You are part of the urban habitat and your presence certainly has some effects, both in helping the spider survive and in making its existence more difficult.

....

Scientists find answers to their questions by making up experiments, comparing the results carefully and then drawing conclusions from what they have seen. The activities offered here are designed to help you understand and participate in the scientific way of observing and learning. Empowered with this method, you can explore the urban landscape without being armed with facts or accompanied by an expert. Finding out about the habits and behavior of your wildlife neighbors can enrich your life in the city and your appreciation of the ecosystem in which you live.

SIX

Wildlife By The Yard:
XERISCAPING FOR WILDLIFE

JIM KNOPF

Along the Front Range of Colorado, gardeners, landscape architects and urban planners have created an eastern urban landscape in the semiarid West by replacing native prairie with an abundance of lush, irrigated lawns, flowers, shrubs and trees. With loving attention and generous quantities of water, the flora introduced from familiar European and eastern North American landscaping became well established. Today, however, residents are paying a high price for this imported landscape style by using increasingly precious water to nourish these water-loving plants. The traditional lawn-dominated landscaping in this region also reduces the rich and lively array of wildlife that can be enjoyed in a more diverse style of landscaping.

There are two gardening strategies that Colorado homeowners and landscapers can employ to enhance the variety of wildlife in developed areas while conserving water in the process. By combining wildlife gardening with xeriscaping (water-efficient landscaping), diverse habitats attractive to urban wildlife can be created. Homeowners often find this type of landscaping less costly and easier to maintain than traditional Kentucky bluegrass lawn-dominated yards.

To a lot of people, landscaping for wildlife means songbirds and butterflies in the garden. For others, it brings to mind skunks, squirrels and raccoons rummaging in the rubbish. Clearly this subject means different things to different people because it involves both attracting beneficial wildlife and, in varying degrees, discouraging wildlife that can damage gardens. Wildlife gardening is a big subject, but there is a wonderful world of discovery for those willing to get involved.

Xeriscape is an equally big subject, and it can be an important part of designing a yard for wildlife. It is not to be confused with cactus-and-gravel gardening. The term for this is ZEROscape. Like wildlife gardening, xeriscape offers tremendous returns. Consider, for example, that 50 percent of the drinking water supplied to cities is typically used for outdoor landscape irrigation, and landscape irrigation can easily be cut in half with xeriscape designs. Xeriscaping can be considered beneficial to wildlife, even without incorporating wildlife gardening principles, because there is less need to disrupt wildlife habitat for water supply development.

Xeriscape Gardening Principles

When setting out to design a xeriscape yard that provides for wildlife, it works best to begin with the seven basic xeriscape principles: (1) plan and design, (2) limit turf and use appropriate grasses, (3) use appropriate plants and zone the landscape, (4) improve the soil, (5) use mulches, (6) irrigate efficiently, and (7) maintain the landscape appropriately. Wildlife gardening strategies can be incorporated into these basic principles.

When designing a garden, cluster elements such as rocks and plants. Varying the sizes of the objects (and the spacing between them) further enhances the results and almost automatically produces visually attractive arrangements.

Limiting turf areas and using appropriate grasses is one of the most effective xeriscape principles because traditional lawns require more water and maintenance than almost any other type of landscaping. It is also an important wildlife strategy, because lawn areas are the least valuable type of landscape for most wildlife. Designing lawn areas to be similar in shape to golf greens and surrounding them with a variety of shrubs, flowers and ground covers is an excellent way to go about reducing lawn areas. The following grasses can be considered in the Denver metro area and along Colorado's Front Range.

- Kentucky bluegrass, currently the most common turf grass used in this area, will grow in either sun or shade, but requires a lot of water—18 gallons per square foot per 20-week season, or .5 inch three times per week during hot weather. Kentucky bluegrass is by far the most water-demanding lawn grass available in Colorado.

- Turf-type tall fescue grasses look like Kentucky bluegrass and grow in sun or shade but require much less water—10 gallons per square foot per 20-week season, or .75 inch once a week in the hottest weather.

- Buffalo and blue grama grasses will grow only in the sun and are soft blue-green in color from about mid-May to mid-September. They require very little irrigation to retain their color in the hottest, driest weather—0 to 3 gallons per square foot per 20-week season, or 0 to .5 inch every two weeks in the driest conditions.

Using appropriate plants and grouping them by water requirements is another very important xeriscape and wildlife strategy. One high-water-demand plant in the midst of otherwise drought-tolerant plants will completely cancel the opportunity to save water and will be a problem if the drought-tolerant plants are intolerant of heavy watering. Mixing redtwig dogwoods and junipers is a good example of what not to do. The dogwoods need lots of water and junipers do not.

By organizing a yard into the following three water zones, great water savings are possible. Because of the diversity that results, a rich wildlife habitat can be created. Typical plants for each zone are listed to help define the zones. Specific water requirements for each zone are the same as those for the grasses, whose water needs are described above.

- High-water-demand zone: Kentucky bluegrass lawn; redtwig dogwood (*Cornus stolonifera*); elderberry (*Sambucus* spp.).

- Moderate-water-demand zone: turf-type tall fescue lawn; potentilla (*Potentilla fruticosa*); common lilac (*Syringa vulgaris*).

- Low-water-demand zone: buffalo and blue grama grass lawn; rabbitbrush (*Chrysothamnus* spp.); junipers (*Juniperus* spp.).

Wildlife Gardening Principles

Keeping the following wildlife gardening principles in mind helps organize the development of a wildlife yard.

Be specific when planning your yard. Most wildlife garden writing is geographically too general to be very helpful when planning a specific yard. For example, mountain ash is often listed as a wonderful plant for birds, but in Denver, few eat its berries. Russian olive, on the other hand, is very popular with almost all local species of wildlife. Incorporate only plants that are useful to wildlife in this geographic area.

List the species you do and do not want to attract, and consider your proximity to the foothills, open space, wild stream corridors or marshes. The distance of your neighborhood from natural habitat makes a difference in the number and kinds of wildlife that will come to your garden. Even in the middle of the Denver metro area, if you are near a pond, you may discover that your backyard goldfish have disappeared down the gullet of a great blue heron. If you live in or near the foothills and mountains, hummingbirds can be lured to your garden if you plant red or orange, tubular, nectar-bearing flowers. If your house is out on the plains, you're less likely to see hummingbirds.

Provide habitat diversity. As you might expect, the greatest habitat diversity—many kinds of trees, shrubs and ground covers at varying heights; a source of water; and proximity to open spaces such as parks, greenbelts and undeveloped land—attracts the greatest variety of wildlife. A study of backyard habitats in suburban Fort Collins revealed that residents of neighborhoods with the five highest habitat diversity ratings reported seeing up to 183 species of birds. Those living in neighborhoods with the five lowest ratings never saw more than 20 species.

Providing plants at low, middle and high levels attracts a variety of wildlife to the different vertical zones. For example, flickers hunt for ants on the lawn, mid-level shrubs provide cottontails with hides from neighborhood cats and dogs, and brown creepers search for insect grubs under bark high on tree trunks. Consciously selecting a variety of sequentially flowering and fruiting plants also will help satisfy a wide variety of species over the entire growing season. Dividing the landscape into different watering zones for greater water efficiency almost guarantees more diversity for wildlife than most traditional yards offer.

Providing edges or borders between plantings and open spaces in your yard can increase the number of animals that spend time there. Wildlife researchers have found that different plant forms—trees, shrubs and flowers—planted around open areas create the desirable "edge effect." This means that more wildlife will be found at the edges of two different habitats than in the center of either one. Hence more wildlife will converge at the intersection of lawn and shrubbery than will be found on pure lawn or in pure shrubbery. Another benefit is that wildlife can be seen better from the edge of a lawn or ground cover area.

Provide for the basic needs of wildlife. The wildlife garden should furnish essential food, water, shelter and safe places in which to reproduce and raise young. Select plants that provide a nutritious diet for wildlife and that protect them from predators and the elements. Include a birdbath or garden pool for drinking water, and provide supplemental feeder offerings and nest boxes.

Food. It is important to select vegetation that produces food year-round for wildlife. For example, fruits of redtwig dogwood, American plum and chokecherry ripen early, are eaten in fall by numerous birds, then drop off. Food plants attractive to various birds that ripen later and hold their fruit throughout winter are Washington hawthorn, firethorn and Rocky Mountain juniper.

A mixture of native and introduced plants that are adapted to local growing conditions offers the greatest wildlife opportunities. In addition to the variety of native plants, introduced plants are very popular with local wildlife because many of them are great fruit and nectar producers.

A wildlife plant list, derived from observations along Colorado's Front Range, can be found at the end of this chapter. It provides a selected list of plants attractive to local urban wildlife and describes how wildlife uses those plants.

When purchasing plants at a nursery, it is helpful to know both the common and scientific names of plants, as well as the cultivar or variety name. If you don't know the name of the plant, try taking a photograph to the nursery of the plant you are interested in purchasing.

Water. A consistent supply of water is essential in any wildlife garden. The results of providing even a simple birdbath can be very dramatic. This is especially true in the semiarid West, where long periods of drought are frequent. The location of the water determines what will be attracted. Placing water on the ground, for example, gives access not only to birds but small mammals as well. Experimenting is valuable. You might try a setup involving running or dripping water, whose sound is very attractive to wildlife.

Shelter and protection. Evergreen shrubs and trees are important in the wildlife yard because of the considerable year-round shelter they offer from predators and winter winds. The variety of cavity-nesting urban birds that can be coaxed to nest in birdhouses if each species' special needs are met is quite impressive. Cavity nesters such as wrens, bluebirds and flickers may need an artificial nest box when aggressive starlings take over their nests, or if the supply of natural cavities is limited. Birdhouses should be placed in appropriate habitat and should be designed to meet the special needs of each bird. For example, house wrens, western bluebirds and mountain bluebirds generally do not use birdhouses in cities, but will on the urban fringe in residential communities. Nest boxes with small entry holes (1 to 1¼ inches in diameter) allow wrens to enter while excluding the larger starlings which require a 2-inch hole for clearance. Northern flickers do nest in Denver and are attracted to nest boxes with sawdust in the bottom.

Protection from danger includes fencing out neighborhood dogs and creating open areas on the ground near bird feeders and birdbaths to safeguard against neighborhood cats. Hummingbird-attracting plants can grow up through chicken wire spread on the ground in spring, and the wire will discourage cats that like to sit in the garden and pounce on hummingbirds. The wildlife repellent Ropel is reported to be quite effective at discouraging cats from hanging out under bird feeders and keeping flickers from drumming on the cedar siding of houses and may also be effective at discouraging cats from sitting among hummingbird plants. Hawk silhouettes and ribbon streamers may reduce collisions of birds with picture windows. Avoiding the use of biocides is critical to the creation of an effective wildlife garden. It is particularly important in protecting insect populations that attract many birds people commonly want. The rapidly increasing supply of effective, environmentally acceptable remedies already offers just about everything any gardener will need. Organic gardening is definitely an important strategy.

Provide for wildlife nuisance control. There are clever and troublesome animals to contend with in the wildlife yard. Accepting the challenge and applying some ingenuity not only makes it possible to minimize most conflicts with wildlife, but also may become an enjoyable and educational undertaking. Here is just one tip of the kind you might discover in your research: Raccoons won't get into a raspberry patch if you scatter mothballs in sandwich bags around the garden.

Black plastic garden netting (available at most area garden stores) used as a perimeter fence often works wonders at preventing deer from browsing garden plants. The deer could easily rip it down, but they rarely do. It is much easier to deal with than chicken wire, and it is nearly invisible from a distance.

Though deer sooner or later sample almost any plant they encounter, there are a few plants that seem to be relatively immune from severe devastation. These include daffodils, iris, yarrow, creeping mahonia, piñon pine and Colorado blue spruce (see Table 2 in Chapter 8). A repellent called Deer Away has definitely kept deer from eating many of their traditionally favorite plants like tulips, crocus and roses.

....

Xeriscape gardening for wildlife is a complex subject. Those gardeners who are willing to become involved in combining these two gardening themes will be well rewarded when their landscapes are transformed into delightful, easily maintained gardens full of mystery and magic. Appreciating a wildlife garden, like appreciating art or music, requires some conscious consideration. Curiosity, simple observation skills and a sense of wonder help a lot. With xeriscaping and wildlife gardening you'll discover there is a lot more going on in your backyard than you ever thought possible.

RICHARD HOLMES

A pair of masked, ring-tailed raccoons scurries up the brick wall of a Boulder home after being caught raiding the cat's dish. These nocturnal mischief-makers are so endearing that such boldness is not only tolerated but often encouraged by humans.

COLORADO'S FRONT RANGE WILDLIFE PLANTS

This list of selected plants and the wildlife species attracted to them is derived from observations in residential settings along Colorado's Front Range. The wildlife species included are only the most interesting or frequent users of the plants in the list.

PLANTS FOR HUMMINGBIRDS

Shrubs
Butterfly bush (*Buddleia* spp.)

Vines
Trumpet creeper (*Campsis radicans)*

Flowers
Scarlet monarda (*Monarda* sp.; red varieties)
Scarlet bugler (*Penstemon barbatus)*
Murray's penstemon (*Penstemon murrayanus)*
Scarlet hedgenettle (*Stachys coccinea)*
Double bubble mint (*Agastache cana)*

PLANTS FOR BUTTERFLIES

Shrubs
Butterfly bush (*Buddleia* spp.)
Bluemist spirea (*Caryopteris* sp.)
Rabbitbrush (*Chrysothamnus* spp.)

Flowers
Gaillardia (*Gaillardia* spp.)
Pitcher sage (*Salvia pitcheri)*
Maximilian's sunflower (*Helianthus maximilianii)*
Russian sage (*Perovskia atriplicifolia)*
Threadleaf groundsel (*Senecio longilobus)*
Purple cone flower (*Echinacea purpurea)*

PLANTS FOR BIRDS, MAMMALS AND INSECTS

Trees

• Domestic apples (*Malus* spp.)
Buds: evening grosbeak, house finch, waxwings
Fruit: fox squirrel, raccoon, striped skunk, red fox
Remarks: Cottontail, deer mouse and house mouse feed on fruit, bark and wood. Bees pollinate flowers.

• Green ash (*Fraxinus pennsylvanica*)
Seeds: blue jay, house finch, evening grosbeak, fox squirrel, house mouse, deer mouse
Buds: evening grosbeak
Remarks: Large-trunked trees provide dens for raccoon.

• Aspen (*Populus tremuloides*)
Buds: house finch, cottontail, squirrels
Catkins: many birds
Associated insects:
many birds
Tree cavities:
used by many birds for nesting
Remarks: Meadow voles nest in aspen root system.

• Boxelder (*Acer negundo*)
Seeds and buds:
evening grosbeak, fox squirrel
Associated insects:
warblers
Remarks: Cavities used by screech owl. Cottontail browses on twigs and bark.

• Cottonwoods (*Populus* spp.)
Nest sites:
American robin, northern oriole, mourning dove, common grackle, northern flicker, squirrels

• Crabapple (*Malus* spp.)
Buds: evening grosbeak
Flowers: house finch
Fruit: house finch, evening grosbeak, blue jay, American robin, waxwings, downy woodpecker, Townsend's solitaire, fox squirrel, raccoon, striped skunk, red fox
Remarks: Cottontail, house mouse and deer mouse eat fruit, bark and wood.

• Hackberry (*Celtis occidentalis*)
Fruit: American robin, northern flicker, rufous-sided towhee, nuthatches, cedar waxwing, fox squirrel, raccoons, striped skunk
Associated insects:
many bird species

• Hawthorns (*Craetagus* spp.)
Cover (thorns repel predators):
many bird species
Fruit: northern flicker, American robin, waxwings, Townsend's solitaire, evening grosbeak, raccoon, fox squirrel
Remarks: Cottontail eats bark and buds from fallen twigs. Bees love flowers.

• Oaks (*Quercus* spp.)
Acorns: blue jay, white-breasted nuthatch, American crow, common grackle, squirrels.
Remarks: Thickets are good cover for rabbits, chipmunks, mice and deer.

• Russian olive (*Elaeagnus angustifolia*)
Seeds: many birds, plus raccoon, squirrels, deer
Fruit: red fox, striped skunk, raccoon, fox squirrel
Remarks: Bees pollinate flowers.

• Douglas fir (*Pseudotsuga menziesii*)
Evergreen cover:
many species of birds and mammals
Seeds: finches, evening grosbeak, fox squirrel, mice
Associated insects:
chickadees, nuthatches, brown creeper, woodpecker
Nest sites in cavities (mostly in foothills):
woodpeckers, chickadees, nuthatches, wrens, bluebirds, swallows
Remarks: Cottontail browses on foliage.

• Ponderosa pine (*Pinus ponderosa*)
Evergreen cover: many species of birds and mammals
Seeds: finches, nuthatches, chickadees, red crossbill, chipmunks, squirrels, mice
Nest sites:
American robin, mourning dove, broad-tailed hummingbird (in foothills)
Remarks: Cottontail browses on fallen needles and branches.

• Colorado blue spruce (*Picea pungens*)
Evergreen cover:
extremely dense cover very important to many wildlife species
Seeds: finches, chickadees, juncos, nuthatches, pine siskin, squirrels, mice
Remarks: Cottontail browses on fallen branches and needles.

Shrubs

• Chokecherry (*Prunus virginia*)
Fruit: waxwings, American robin, northern flicker, house finch, blue jay, rufous-sided towhee, raccoon, house mouse, deer mouse, red fox, striped skunk, fox squirrel
Remarks: Cottontail browses on fallen twigs. Bees pollinate flowers.

• Rocky Mountain juniper (*Juniperus scopulorum*)
Evergreen cover:
most species
Seeds: Townsend's solitaire, American robin, waxwings, mourning dove, evening grosbeak, chickadee
Fruit: deer mouse, house mouse
Nest sites:
many species, including deer mouse

• Scrub oak (*Quercus gambelii*)
Acorns: scrub jay (along foothills), band-tailed pigeon
Associated insects:
chickadees, rufous-sided towhee
Remarks: White-breasted nuthatch, American crow and common grackle use other seeds that collect in thickets of scrub oaks.

• Piñon pine (*Pinus edulis*)
Evergreen cover: many species
Seeds: scrub jay (along base of foothills)

• Wild plum (*Prunus americana*)
Cover: thickets good protection for many species
Fruit: northern flicker, American robin, house finch, blue jay, chipmunks, squirrels, raccoon, striped skunk, red fox
Remarks: Bees are attracted to sweet-smelling flowers.

• Woods rose (*Rosa woodsii*)
Fruit: waxwings, tree sparrow, American robin, Townsend's solitaire, raccoon
Remarks: Cottontail feeds on bark and buds of twigs. Bees favor flowers.

• Rabbitbrush (*Chrysothamnus* spp.)
Flowers: Butterflies, bees
Associated insects: many bird species

• Nannyberry (*Viburnum lentago*)
Fruit: American robin, northern flicker, red fox, fox squirrel, raccoon
Remarks: Cottontail feeds on fruit and bark. Bees pollinate flowers.

• Firethorn or Pyracantha (*Pyracantha* spp.)
Cover: semi-evergreen, dense, thorny branches good wildlife cover, especially in winter.
Fruit: waxwings, house sparrow, blue jay, white-crowned sparrow, American robin

• Oregon holly-grape (*Mahonia aquifolium*)
Evergreen cover: good for many species
Fruit: American robin, fox squirrel, raccoon, mice, cottontail

• Highbush cranberry (*Viburnum opulus*)
Fruit: American robin (occasionally get drunk on fermented berries)

• Elderberry (*Sambucus* spp.)
Fruit: American robin, northern flicker, blue jay, waxwings, house finch, mourning dove, white-crowned sparrow, rufous-sided towhee, raccoon, red fox, fox squirrel, striped skunk, mice
Remarks: Migrating fall birds make use of berries. Bees pollinate flowers.

• Rocky Mountain sumac (*Rhus glabra cismontana*)
Seeds: downy woodpecker, northern flicker, Steller's jay

• Japanese barberry (*Berberis thunbergii*)
Cover (dense branching and thorns repel predators): many species
Seeds: evening grosbeak, American robin, waxwing

•Pin cherry (*Prunus pennsylvanica)*
Fruit: evening grosbeak, American robin, waxwings, American crow, blue jay, downy woodpecker, northern flicker, house finch, Townsend's solitaire, migrating fall birds, house mouse, deer mouse, raccoon, fox squirrel, striped skunk, red fox
Remarks: Cottontail browses on fallen twigs.

Vines

•Virginia creeper (*Parthenocissus quinquefolia*)
Fruit: blue jay, northern flicker, house finch, house sparrow, American robin, chickadees, downy and hairy woodpecker, raccoon, fox squirrel, mice

•Grape vines (*Vitis* spp.)
Fruit: American robin, northern flicker, northern oriole, western tanager, squirrel, raccoon, fox squirrel, red fox, striped skunk

Flowers

•Butterfly weed (*Asclepias tuberosa)*
Nectar source for many butterflies, including monarch and black swallowtail. Hummingbirds, bees and other insects attracted to plant. Butterfly caterpillars feed on plant.

•Clovers (*Trifolium* spp.)
Foliage eaten by cottontail and thirteen-lined ground squirrel. Great Plains checkerspot butterfly feeds on nectar. Caterpillar food for clouded sulphur. Flowers desired by bees and butterflies. An area in one's yard set aside just for clover is beneficial for wildlife.

•Common sunflower (*Helianthus annuus)*
Superb plant for wildlife. Mice, fox squirrel and thirteen-lined ground squirrel eat seeds. Painted lady and Great Plains checkerspot butterfly caterpillars feed on plant. Bees, boxelder bugs and many other insects attracted to flowers. Red-winged blackbird, Brewer's blackbird, brown-headed cowbird, mourning dove, house finch, pine siskin, downy woodpecker, white-breasted nuthatch, black-capped chickadee, dark-eyed junco, house sparrow and white-crowned sparrow feed on sunflower seeds.

•Evening primrose (*Oenothera strigosa)*
Finches and black-capped chickadee eat seeds. Nectar favored by spinx moth, bees and other insects. Can be invasive if not managed. Tall-stemmed varieties best.

•Goldenrod (*Solidago canadensis, S. missouriensis, S. speciosa)*
Flowers lure bees and butterflies such as painted lady, monarch and green fritillary. Pine siskin and other finches feed on seeds. Cottontail attracted to foliage. Plant rarely causes hay fever.

•Marigold (*Tagetes erecta)*
Bees and butterflies such as clouded sulfur attracted to flowers. Hummingbirds infrequently come to plant. House finch and house sparrow eat seeds. Marigolds good for repelling certain pests of other garden plants.

•Mullein (*Verbascum thapsus)*
Northern flicker, black-capped chickadee, downy woodpecker and finches feed on seeds. Insects attracted to flowers. Mullein can be invasive if not managed.

•New England Aster (*Aster novae-angliae)*
Nectar is source for honeybees and many butterflies such as checkered skipper and clouded sulfur. Birds feed on seeds. Cottontail favors foliage. Many kinds of asters provide nesting material for spring birds. Seed heads produced in fall stay on plant into spring. Seed heads used by mice for nesting.

•Pink bergamot (*Monarda fistulosa)*
Nectar loved by bees, hummingbirds and butterflies such as monarch and green fritillary. Pine siskin visits plant for seeds.

•Showy milkweed (*Asclepias speciosa)*
Caterpillar food for monarch butterfly. Many insects, butterflies and bumblebees attracted to flowers. Fluffy seeds used by late-nesting birds.

•Teasel (*Dipsacus sylvestris)*
Pine siskin and other finches eat seeds. Insects and bees attracted to flowers.

•Zinnia (*Zinnia elegans)*
Sparrows and finches eat seeds. Many butterflies such as monarch, western tiger swallowtail, orange sulfur, painted lady and silver-spotted skipper feed on nectar. This species is the prime zinnia for wildlife in Colorado.

....

The list of plants for hummingbirds and for butterflies was prepared by Jim Knopf. The list of plants for birds, mammals and insects was prepared by Tina Jones.

SOURCES

Damrosch, B. 1982. Theme Gardens. Workman Publishing, New York, 224 pp.

Hammerson, G. A. 1982. Amphibians and Reptiles in Colorado. Colorado Division of Wildlife, Denver, 131 pp.

Henderson, C. L. 1981. Landscaping for Wildlife. Minnesota Dept. of Natural Resources, St. Paul, 144 pp.

Henderson, C. L. 1981. Woodworking for Wildlife. Minnesota Dept. of Natural Resources, St. Paul, 47 pp.

Wells, M. 1988. Classic Architectural Birdhouses and Feeders. Malcolm Wells, 673 Satucket Road, Brewster, Massachusetts, 92 pp.

Jansen, C. 1974. Backyard Wildlife. Colorado Outdoors 23: 31-34.

Merilees, W. J. 1989. Attracting Backyard Wildlife. Voyageur Press, Stillwater, Minnesota, 159 pp.

Tekulsky, M. 1985. The Butterfly Garden. Harvard Common Press, Boston, 144 pp.

Tufts, C. 1988. The Backyard Naturalist. National Wildlife Federation, Washington, D. C., 79 pp.

SEVEN

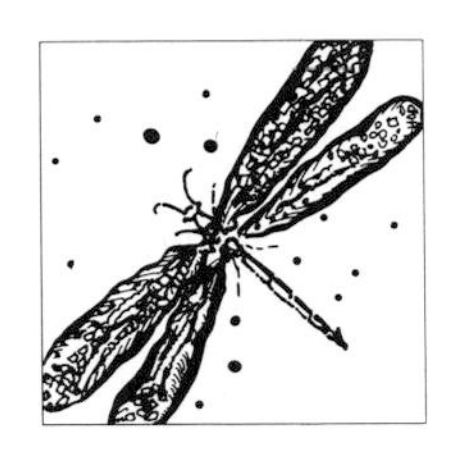

Wildlife Watching
FOR INSOMNIACS

ELIZABETH A. WEBB

Bracketing the work hours, from dusk to dawn, there exists a subtle, little-known night shift of nonhuman urban bustle. Along the Platte River Greenway, black-crowned night-herons quietly stalk suckers and minnows around shallow pools in the cool of a summer sunrise. Yellow-flowered prickly lettuce wedges its way into antique sandstone sidewalk cracks at Sixteenth and Pearl during a twilight rain shower. A monarch butterfly alights on showy milkweed to lay her eggs in the midst of evening rush hour traffic. Dusk brings out high-flying common nighthawks which, like winged Dustbusters, scoop up aerial insects over the United Bank Building. After nightfall, red foxes hunt meadow voles in the lush willows and sweet clover that line reclaimed gravel pits along Clear Creek at Interstate 25. A lone, inconspicuous Woodhouse's toad makes itself at home on someone's front door welcome mat, gulping stunned bugs fallen from the searing porch light overhead.

As street lamps all around Denver automatically turn on at dusk, so do nocturnal insects. Among the girders and reflective glass of downtown, night revelers are in evidence, and so is the calming influence of urban wildlife.

You don't have to pack up the kids in the four-wheel drive to head for wilderness. Instead, stay close to home and try an attitude adjustment. Explore wildlife that you have previously stepped on, swatted, reviled unfairly, or simply overlooked in the urban setting. During morning rush hour, spend 90 seconds idling at a clogged intersection watching iridescent barn swallows pick off bugs while careening through stopped traffic. Intersections are patches of urban open space that provide insect-hawking and nesting sites for these swallows. Birds are here because bugs are here for the eating in large numbers.

The bugs' presence may be the result of what atmospheric scientists call the boundary layer effect. Insects become sandwiched between a layer of turbulent air stirred up by moving traffic at ground level and a layer of still air above. When cars stop at a red light, there is a momentary opportunity for swallows to sweep the intersection of trapped bugs. Barn swallows are most evident at intersections on cool, damp mornings during the late spring and early summer. Hot exhaust rising from idling cars can raise the air temperature at intersections around 2 to 3 degrees F (1 to 1.5 degrees C) above surrounding areas. Barn swallows, being temperature sensitive, may be attracted by the warmer air. They have been rumored to build their mud nests under traffic light hoods. If this is the case, they needn't go far to eat out.

Some of the most desirable wildlife is found at the most undesirable places and at the most unforgiving hours. Derelict industrial sites, fetid pools of water, railroad yards, garbage dumps and crumbling parking lots have their own strange brew of natural history. Some wildlife species still inhabit Denver's remnant patches of natural features—short-grass prairie, cottonwood-willow river bottom and cattail marsh. Others live in human-built habitats such as backyard gardens, chimneys, bridge abutments, skyscraper ledges and church belfries.

For insomniacs, easy-to-observe nocturnal insects hang out at all-night gas stations, automatic teller machines, phone booths, 7-Elevens and night-lit car lots. One can compare the relative entomological merits of these cultural icons of the night through simple observation. At all-night gas stations, sinewy crane flies that resemble overgrown mosquitoes do push-ups to keep warm on the perpetually frigid lavatory walls. Minute silvery-green phantom midges dart in and out of the lights; care must be taken not to confuse them with Day-Glo tennis ball fuzz. On an evening field trip to a Conoco station, friends and I found army worms, click beetles, boxelder bugs, black field crickets, green lacewings, sowbugs, and leafhoppers on the well-lit garage walls. However, the wonder of urban wildlife is not apparent to all. While we were inspecting some flattened fauna on the massive chrome grille of a pickup parked at the pump, our forceps and collecting jars in hand, a mean-looking cowboy returned to his truck and asked incredulously, "Whaddre you boys doin?" We rounded up our gear and made a fast getaway.

It is a well-known phenomenon that nocturnal insects are attracted to light. Their bag of tricks includes use of celestial cues to locate resources such as mates and food. An insect needing to fly in a straight line at night probably can do so best by orienting at a constant angle to a distant light landmark such as the moon, stars or illuminated clouds. If the light source is a street lamp, porch light, bug zapper or billboard in the near distance, moths flying toward it may make a spiral, zigzag or circular pattern. Or they may make a straight shot for it, sometimes crashing into the light.

Urban wildlife enthusiasts must take particular pains to appear casual when bug-watching at all-night automatic teller machines and phone booths so as not to be mistaken for bank robbers or other felons. If you look carefully enough for insects, however, you may make a haul despite your innocence. Our observant group collected four cents on a single outing. The insect population is higher inside ATMs and phone booths than outside for three reasons. Inside they are better lit, there is human flesh in a holding pattern, and there is no wind. Most insects cannot fly in even a slight breeze. To tiny creatures, a zephyr is like a tornado and a raindrop like a bombshell.

Most 7-Elevens in the Denver metro area have the Plus System 24-hour banking service, so one can indulge in the pleasures of an all-night convenience store and an ATM with a single stop. One night while perusing the bright white walls of a 7-Eleven, I noticed remnants of a tubular mud dauber wasp nest affixed to a natural section of moss rock. The nest had been destroyed, probably by the manager. This is regrettable—mud daubers are great to have around a food store because they provision their nests with captured spiders that would otherwise be cluttering the display windows with untidy webs.

Night-flying insects are attracted to buildings with bright lights and white walls that reflect light. Midnight shoppers can make the wait in line at the cash register productive by leaving the headlights of their parked cars directed at outside white walls. This will attract swarms of nocturnal insects in the beams and drain the car's battery. When conducting this experiment, it is best not to wear white clothing.

Car lots are fine places to view nocturnal wildlife, as long as no one tries to sell you a car. The glaring overhead lights act as a beacon guidance system that lures night-flying insects. One night in Boulder, a magnificent male Sphinx moth appeared suddenly out of the darkness in a downward arc of bright light at a Mercedes-Benz dealership. He spiraled gracefully upward in a stellar performance, perhaps in pursuit of a pheromone plume released upwind by a virgin female. Adult butterflies and moths live for a relatively brief time, so there is a rush to find a mate.

Like the Mercedes models on the showroom floor, Sphinx moths are classy, streamlined and elegant. During the day, they rest immobile and unseen. With bark-colored forewings neatly folded back over pink-hued hindwings, they blend into their daytime habitat. Large-bodied, they are often mistaken for hummingbirds because they hover while nectaring at flowers. They sip nectar through an unfurled proboscis that looks like a child's flexible straw. Unlikely though it may seem, the dreaded tomato hornworm caterpillar is the larval stage of one species of Sphinx moth.

Sphinx moths are considered to be smart insects because they possess a memory. Over time, they learn not to be fooled by artificial lights. Once they know where to locate resources in a given area by following odors, landscape contours, wind and other cues, they abandon reliance on landmark cues. Only newly emerged, inexperienced adults in mint condition congregate at night lights. The old, torn and faded moths have learned to stay away.

The city at night is a heavily illuminated environment. Outdoor lighting in metropolitan areas has a profound effect on the behavior of nocturnal insects. Studies have shown that light pollution disturbs the flight, vision, orientation, natural movements, reproductive behavior, feeding and protective camouflage of moths. Nocturnal insects are attracted to light at the short end of the electromagnetic spectrum, in the ultraviolet and blue region, because it is the most visible to them. Yellow, orange and red light, at the opposite end of the spectrum, suppresses flight-to-light behavior. This is why yellow porch lights don't attract insects, and why not even the most lowly bugs appreciate the golden arches of McDonald's.

City street lamps are the main source of confusion for nocturnal insects because there are so many to attract them. Tungsten filament, low-pressure sodium, high-pressure sodium, metal halide, mercury, incandescent and fluorescent lights contribute to city glow. Each type has its own spectral properties. Vis-à-vis insects, there are good ones and bad ones.

The most commonly used street lighting in Denver today is the high-pressure sodium lamp, which has, for the most part, replaced the older mercury lamps. This type puts out less blue light than mercury lamps but is still spectrally broad and thus attracts insects. The low-pressure sodium lamp, on the other hand, is spectrally narrow in that it puts out an orange-yellow light and almost no blue. The human eye is particularly sensitive to this narrow band of orange-yellow light, so that less energy is required for it to produce effective city lights. A revealing comparison can be made between these two types of lighting at McNichols Arena. The pale lower wall has a row of high-pressure sodium vapor lights that attract such beauties as green lacewings, caddisflies, ground beetles, midges, winged ants, milkweed bugs, cutworm moths and elm beetles. Most of these insects fly in from adjacent landscaping on the upper west side of the arena. Dotting the parking lots surrounding Big Mac are the orange-tinted low-pressure sodium vapor lights that attract nothing, using less energy and not interfering with urban wildlife.

Optimal conditions for viewing nocturnal urban insects can be found on cloudy, moonless nights. A full moon presents a landmark that competes with artificial lights. An isolated source of artificial light is more likely to attract moths than high-density lighting, which may even have a repellant effect. This effect was verified by a field trip to Mile High (1.6 km) Stadium during a preseason Broncos night game. There were 76,247 lunatic fans in attendance, but not a single night-flying insect. The large and competing banks of lighting apparently lessened the attraction of insects to the stadium. Sports fans will note that this phenomenon is known as lunar phobia.

Bug zappers, on the other hand, have the opposite effect. Bug zappers are solitary blacklight traps that emit utterly irresistible ultraviolet light and fry the insects that come into contact with an outer electrocuting grid. Nocturnal insects swarm to them. This is ironic, since zappers are meant to reduce residential insect populations. Neighbors of mine put a bug zapper on their back deck for several summers. Every few minutes, a sharp sizzle would break the evening's calm and, undoubtedly, two satisfied consumers would reflect on the wisdom of their purchase. Unbeknownst to them, the insects probably flew out of their way, perhaps from great distances, to the artificial light. And because most insects land in the vicinity of a bright light without coming into contact with it, the majority of my neighbors' visitors remained unzapped. The net effect of the bug zapper was to dramatically increase the population of nocturnal insects in the neighborhood—to my pleasure. Population increase might not always be the case, however. It is possible that bug zappers set up in environmentally sensitive areas could obliterate small, localized colonies of rare insects or hinder establishment of interesting species. Zappers also kill insect predators such as dragonflies, which are a natural control for the likes of mosquitoes.

The laws of ecology are not repealed at city borders. The burgeoning field of street lamp ecology is a complex web of predator-prey relationships. Spiders, toads, nighthawks, bats and skunks, all nocturnal predators, hunt insects gathered at street lamps. Predation is rampant because insects are easy targets when congregated at lights.

A walk around City Park Lake at dusk will reveal common nighthawks buzzing over the water and adjacent street lamps in search of flying bugs. Nighthawks have tiny bills and cavernous gapes, which means that when they open their outwardly unassuming mouths they look as if they could suck in a small elephant. Forward-slanting bristly feathers line the bill to sweep insects into the mouth like an animated broom. This further increases the ample capacity of the birds' aerial insect-catching equipment.

Nighthawks are members of the goatsucker family, but they refrain from sucking in goats. The name "nighthawk" is, by comparison, only a slight misnomer. Nighthawks are not truly nocturnal, nor are they hawks. They hunt mostly during the crepuscular hours. Like falcons and swifts, which are also aerial avian predators, they have tapered, backward-pointed wings designed for speed and high maneuverability in flight for pursuing darting prey.

Nighthawks might be overlooked in the city were it not for their piercing flight calls, which sound like noises made by raspy squeeze toys. Birds are legendary for their songs and calls as a means of communication. They also communicate using nonvocal acoustical signals such as bill-clacking, drumming, and wing-whistling. Male common nighthawks produce a hollow booming sound when air passes through their vibrating primary wing feathers during swooping courtship displays. The aerial displays and booming extend through the summer nesting season. Nighthawks do not build nests. Instead they deposit their gravel-colored eggs on the open ground or on skyscraper rooftops—substrates similar in anatomy. Thus common nighthawks have made a successful transition to the urban environment. Denver's rooftop-reared birds—the ground-nesting nighthawks and killdeers, as well as cliff-nesting peregrine falcons—will be the focus of increased attention as more and more wild nesting sites become urbanized. The staying power of many wild birds in Colorado will depend on their ability to nest in human-created habitats.

Big brown bats hunt night-flying beetles over city street lamps. The night provides the bats with a seemingly limitless source of coleopteran cuisine—beetles that include scarabs and long-horned borers, common night-flyers attracted to lights. I would guess that from a bat's perspective, they are big, juicy and pleasant-tasting—the perfect bat meal.

Big brown bats begin to stir in their day roosts before sunset, testing for light levels. They emerge ready to eat shortly after sunset. Bats can be mistaken for nighthawks, but their flight is more erratic. Using their high-tech echolocation system, they bounce high-frequency pulses off minuscule insects, and the pulses bounce back to their oversized ears in the darkness. Gram for gram, watt for watt and dollar for dollar, the bat sonar system is billions of times more efficient than what the military-industrial complex has developed. Some bats have been observed catching as many as 600 mosquitoes per hour.

Moving in on an outmaneuvered beetle, a big brown bat seeks a direct hit into the mouth. It may catch large prey in a wingtip or uropatagium (elastic tail membrane) and transfer it to the mouth. To do this, it does a split-second barrel roll and tumbles down with a slight loss of altitude while tucking in the wing or tail. Flips, rolls and rapid changes in trajectory to catch evasive insects are the signature of bats' erratic, fluttery flight silhouetted against a dimming sky.

Most of us don't realize how close to home big brown bats live. And when we do, we panic for no reason. Big brown bats do not suck blood, attack people or get tangled in hair. They are poor transmitters of rabies. The last human case of rabies in Colorado was in 1931—it was caused by a dog bite. These unloved mammals have the innocent faces of puppy dogs and beautiful copper-colored fur. Furthermore, bats are important indicators of a healthful human environment. Like caged canaries in a mine shaft, they can give advance warning of toxic contaminants in our neighborhoods. They should be welcome neighbors.

Big brown bats roost in buildings, tree holes and rock fissures, under loose bark and between expansion joints of bridges. In buildings, they roost by hanging upside-down all day in chimney chases, between walls, behind shutters, in attics, under eaves or in other vacant spaces that are fairly closed and dark. Big brown bats are quiet creatures, live in small colonies, and leave only subtle signs of their presence. They evacuate body wastes after night forays just before reentering the roost, leaving a small white guano stain outside. Big brown bats forage over city parks, cemeteries, lakes, waterways and older residential neighborhoods. Newer homes allow bats less access because they are built more tightly for energy conservation. Tight construction prevents air infiltration and therefore excludes bats. New homes also often have soffits, window caulking and fewer access areas such as wall cracks through which to enter and establish a day roost. Even so, bats come into living quarters of homes by predictable routes—through open windows and doors, ungrated chimneys and tattered screens. Although we humans are the most destructive creatures on the planet, it can be said in our defense that we have created habitat for bats to expand into urban centers.

On a twilight walk along the Highline Canal bike path at Fairmount Cemetery, foraging big brown bats were so dense that they collectively squeaked to avoid aerial collisions. A red fox appeared, ever so splendid, from behind a marble headstone. The sweet but admittedly nauseating smell of a striped skunk's spray wafted across the bike path, although the perpetrator itself did not. Certain mammal observations are more fleeting than others. I find that in encounters with what others call problem or pestiferous wildlife, it helps to understand the behavior behind the perceived nuisance. The striped skunk does not exist solely to annoy people with its pungent musk. The skunk is far more vulnerable than humans, and it knows this. To protect itself, it patiently waves its bushy tail and stamps its feet first as a warning. When this fails to deter the intruder, it turns around and sprays on target an oily, yellow mist produced by glands under the tail. This defense allows the skunk to wander through city neighborhoods with impunity.

Darkling beetles emerged from under rock hides to crisscross the bike path, doing what they do best on a Saturday night—eating or being eaten. They often turn up as undigested pieces of chitin under street lamps in the scat of Woodhouse's toads. (One can tell a lot about an animal by looking at its scat—possibly more than one cares to know.) Knotweed, the ultimate sidewalk crack plant, was growing prostrate on the pavement to withstand constant trampling by feet, paws and bike tires.

Lining the pavement were shoulder-high Russian thistles, noxious weed or glorious wildlife plant, depending on your point of view. When wearing Velcro-festooned sneakers on an urban nature walk, never leave the confines of the path unless you wish to be dressed as a Russian thistle for the rest of your life.

At today's supersonic pace, the pressures and anxieties of urban living underscore the need for regular contact with nature. If you find yourself unable to sleep at night, don't automatically flip on the TV—flip on the porch light instead.

SOURCES

Adams, L. W., and D. L. Leedy, eds. 1987. Integrating Man and Nature in the Metropolitan Environment. National Institute for Urban Wildlife, Columbia, Maryland, 249 pp.

Colorado Division of Wildlife. 1984. The Bats of Colorado: Shadows in the Night. Colorado Division of Wildlife, Denver, 23 pp.

Ehrlich, P. R., D. S. Dobkin, and D. Wheye. 1988. The Birder's Handbook: A Field Guide to the Natural History of North American Birds. Simon and Schuster, New York, 785 pp.

Frank, K. D. 1988. Impact of outdoor lighting on moths: An assessment. Journal of the Lepidopterists Society 42(2): 63-93.

Goode, D. 1986. Wild in London. Michael Joseph Ltd., London, 186 pp.

Janzen, D. H. 1984. Two ways to be a tropical big moth: Santa Rosa saturniids and sphingids. Oxford Surveys in Evolutionary Biology 1: 85-140

Jones, J. K., Jr., D. M. Armstrong, R. S. Hoffmann, and C. Jones. 1983. Mammals of the Northern Great Plains. University of Nebraska Press, Lincoln, 379 pp.

Phillips, G. L. 1966. Ecology of the Big Brown Bat (Chiroptera: Vespertilionidae) in northeastern Kansas. American Midland Naturalist 75(1): 168-198.

Tuttle, M. D. 1988. America's Neighborhood Bats. University of Texas Press, Austin, 96 pp.

ACKNOWLEDGMENTS

I would like to thank the following friends for their assistance: Dr. Ric Peigler, Mike Weissmann, Wendy Shattil, Bob Rozinski, Dr. Carron Meaney, Dr. Alex Cringan, Steve Bissell, Camille and Cynthia Tchilinguirian, Laurie Kuelthau and Joanne Carter.

PHOTOGRAPH LEGENDS

PAGE 93

A house sparrow checks himself in a rear-view mirror in a Littleton parking lot. He waited long enough for the photographer to run to her car, grab her camera and record the moment. Many photos in this book are equally serendipitous.

PAGE 94 ABOVE

A meadowlark's shadow falls on a land developer's sign. Conversely, encroaching development is falling fast on wildlife on Denver's outskirts.

PAGE 94 BELOW

A female mallard is hopelessly entangled around the neck and between the mandibles by plastic rings from a sixpack of beverage cans. The bird eluded the photographer's rescue attempt. It probably starved or strangled later. Plastic rings should be cut with scissors before they are thrown away.

PAGE 95 ABOVE

Burrowing owls perch on a sign near Arapahoe Road and Peoria Street in South Denver shortly after sunrise. Menacing glares lend authority to the posted warning. The entrance to their burrow is nearby, in a prairie dog colony.

PAGE 95 BELOW

A male red-winged blackbird pauses in Cherry Hills Village.

PAGE 96

A female mallard leaps into a nearly empty swimming pool near Golden. She visited the pool often to take baths and look for food in the bit of rainwater in the bottom. The owners of the pool named her Gertrude. Mallards are the most common ducks in the Denver area.

PAGE 97 ABOVE

Bikers and walkers along the South Platte River through Denver should see many black-crowned night-herons in summer. These patient fishermen wait where the water is turbulent: below dams and spillways, by storm sewers and along rapids.

PAGE 97 BELOW

Although birds can't read, this goose-stepping parade is headed in the right direction nevertheless. Imagine the disappointment of these Canada geese when they found the swimming pool at Rocky Mountain Arsenal dry and abandoned.

PAGE 98

In winter, Viele Lake in Boulder is alive with waterfowl. A bubbler in the middle of the lake and the birds themselves keep part of the water ice free. Hooded mergansers, wood ducks, greater white-fronted geese and double-crested cormorants are only a few of the species found here.

PAGE 99 ABOVE

Pronghorn graze east of Aurora at the Plains Conservation Center. Keen eyesight and speed are their main defenses against attack, so they favor the wide open grasslands of the high plains. Pronghorn presently live on the south and east outskirts of Denver.

PAGE 99 BELOW

American coots cruise on a pond at Rocky Mountain Arsenal as a 727 makes its final approach to Stapleton International Airport. Stapleton adjoins the arsenal on the south.

PAGE 100 ABOVE

The first rays of sunlight impart a golden aura to a northern bobwhite at Rocky Mountain Arsenal. Although these quail are native to southeastern Colorado, this one more likely descended from birds released years ago at the arsenal for hunting and for field trials of sporting dogs.

PAGE 100 BELOW

Denverites share a blazing sunset at City Park Lake.

SUSAN WORTH JENKINS

CATHERINE BARNES

J. B. HAYES

J. B. HAYES

GARY A. HAINES

KENT CHOUN

HAROLD ARNOLD

WENDY SHATTIL/BOB ROZINSKI

JOHN B. WELLER

AL WALLS

WENDY SHATTIL/BOB ROZINSKI

WENDY SHATTIL/BOB ROZINSKI

SCOTT M. GILMORE

EIGHT

Boulder's Deer Population

CHARLES H. SOUTHWICK, BRIAN PECK, ANTHONY TURRINI
AND HEATHER SOUTHWICK

One Sunday afternoon in early October, a minidrama unfolded that characterizes much of Boulder's dilemma with deer. Sitting on the north edge of Table Mesa, two of us looked down upon the houses along Kohler Drive. Five handsome mule deer bucks, not yet engrossed in the approaching rutting season, were milling along the edge of a fence encircling a beautiful garden and patio. Inside the house the resident was relaxing in front of his television set, probably watching a football game. With impressive ease, the deer bounded one by one over the wooden fence, which appeared to be nearly five feet high, and began munching on the shrubbery and plants of the garden. With obvious irritation, the homeowner burst through the patio doors, arms flailing, and scared away the intruders. They cleared the fence handily in the opposite direction. The homeowner returned to his TV, and the deer loitered outside the fence looking like five mischievous schoolboys. Within minutes they were back inside the fence chomping again on the garden. The homeowner stormed out a second time. In the next hour, we saw this episode repeated five times. The homeowner's quiet afternoon was thoroughly disrupted, and his plants got a trimming he surely had never intended.

Almost everyone agrees there is a "deer problem" in Boulder, but the approaches to this "problem" vary greatly. Answers to a recent public opinion questionnaire in west Boulder showed that 99 percent of 125 respondents had had deer visit their yards, and 85 percent reported property damage. But only 47 percent felt any action should be taken against the deer; the majority enjoyed seeing them despite the damage they caused. Many people in Boulder consider the deer a source of enjoyment and a part of Boulder's unique natural environment, an environment to be protected and savored. These people feel fortunate to have the deer close at hand, unafraid, adjusting to the urban habitat. But for those whose favorite trees, shrubbery and flowers are devastated by deer, or those injured in automobile-deer collisions on city streets, the deer are out of place and a dangerous nuisance. These people feel the city is negligent in not managing the deer problem more successfully.

There have even been letters to the editor in local newspapers referring to the "spread of disease" by deer, although it is not clear what this means in Boulder. In the eastern and north-central United States, Lyme's disease can be transmitted by the deer tick *Ixodes dammini,* but fortunately, our mule deer do not have this tick. Mule deer in Colorado have another species of tick, *Dermacentor albipictus,* which is not known as a vector of Lyme's disease. Dr. Carl Mitchell, an internationally known medical entomologist in Fort Collins, points out that the transmission of Lyme's disease is not a likely event in Colorado unless there should be a major change in the ecology of the spirochaete bacterium that causes the disease or a major change in the tick populations on deer and small mammals. Furthermore, the real reservoirs of Lyme's disease in states where it is common are small mammals such as wood mice, voles, opossums and raccoons, despite the public impression of deer as the main source. Of two known cases of Lyme's disease in Colorado as of 1989, one was acquired out of state, and the other one, from the Four Corners area of southwestern Colorado, was of unknown origin.

The deer of Boulder are a public issue, often discussed in newspapers and the broadcast media. Despite their high visibility, however, very little scientific literature on their ecology and behavior has appeared. Amid the emotions, it is necessary and helpful to look at the facts as much as possible in an effort to evaluate the pros and cons of Boulder deer. What do we know about these fascinating animals who have chosen to live in our midst?

Boulder's Two Species of Deer

Two species of deer live in and around Boulder. The great majority are mule deer, *Odocoileus hemionus,* a species characteristic of western grasslands and ponderosa pine forests, that is equally at home in the mountains and on the prairies wherever good forage and some trees provide food and cover. One notable quality is their ability to adapt to human presence and live in close association with our homes, our automobiles and ourselves.

The second species, the white-tailed deer, *Odocoileus virginianus,* is less common and much less frequently seen. These deer are shy of people and more secretive in their habits. In Boulder, white-tailed deer inhabit creek bottoms where brush is thick, and they also move up into the ponderosa pine forests of the foothills. They are most easily seen at the southern edge of town toward South Boulder Creek as it emerges from Eldorado Canyon. In this area, between 5 and 10 percent of the deer seen may be white-tails if their favored habitats are watched quietly. White-tails may also be seen northeast of Boulder in the White Rocks area. Michael Sisk, a pre-med student at the University of Colorado who did a comparative study of mule and white-tailed deer in the spring of 1989, observed that the average flight distance of white-tailed deer was 60 yards, whereas mule deer could usually be approached within 5 yards, and Steve Lipshur, a CU biology major who studied mule deer in north Boulder, had deer come up to him and lick his hand. Apparently these deer had received handouts of food from other people.

We have not seen white-tailed deer in north Boulder nor in the central part of the city where mule deer are abundant. In south Boulder, extending toward Eldorado Springs, white-tails can be seen, but it is not always immediately obvious whether one is seeing mule deer or white-tails. The chances are at least 9 to 1, however, that any deer seen are mule deer. Table 1 lists a few of the important distinctions between the two species, but these differences are sometimes blurred because hybrids are known to occur. Hybrid individuals often have intermediate traits; that is, a combination of traits, or a blending of them. Although both species are native to this region, under normal circumstances differences in their habitat preferences and behavior tend to keep them apart. The presence of hybrid animals may indicate habitat changes and ecological disturbances that have broken down the natural barriers between species. Interbreeding might also be the result of a very low white-tail population seeking mule deer mates because members of their own species are difficult to find with regularity.

Population Trends

One of the major issues regarding Boulder's deer population, especially the mule deer, is that of population trends. The conventional wisdom is that the mule deer population has been increasing dramatically, perhaps 10 percent per year, according to newspaper accounts. To address this issue the City of Boulder commissioned a series of ecological studies on the deer, beginning in 1983, to estimate the population and evaluate trends. Boulder Mountain Parks and the Open Space Program have headed this research effort with the cooperation of the Colorado Division of Wildlife staff, who assisted in trapping and tagging the deer. The two city departments asked Western Resource Development of Boulder to initiate the original study in 1983 and 1984.

The annual censuses have been based on a standard technique in wildlife research and management known as the Lincoln-Peterson Index. In this method, a known number of animals in the population are trapped, tagged and released. On the assumptions that these tagged animals are representative of the total population, that they mix randomly with the population, and that their probability of being sighted is the same as that of untagged members of the population, the population is then censused visually by systematic walking surveys that cover the habitat as thoroughly as possible. The ratio of tagged to untagged deer observed can then be used to estimate the total population. When this is done several times by trained field workers, a reasonable estimate of the population may be obtained.

In simple terms, if a total of 100 deer are tagged, and on a subsequent census 100 deer are seen, only 10 of which are tagged, the assumption is that the total population is 10 times the tagged population, or 1000 deer. Once several separate censuses and population estimates have been made, the average is taken, and the standard error of the average can be calculated. This permits confidence limits or reliability estimates to be placed on the population counts.

Census data since 1983 show that the mule deer population as seen in Boulder has indeed increased if the averages are compared. However, the numbers may now be leveling off (Fig. 1). An estimated population of less than 800 deer in 1983 increased to 1100 by 1988 but has since declined to 1007 in 1989. These estimates have standard errors of 3.0 to 8.4 percent, and 95 percent confidence intervals indicate that the average estimates may be in error by as much as 10 percent.

City Deer vs. Country Deer

Do these population estimates represent true growth of the population, or merely the tendency of more deer to come into the city and suburban environment and thus be more readily seen and counted? This is a difficult question to evaluate, and one for which no firm or conclusive answer is available.

One way to explore this question is to compare census counts from limited areas; for example, an area of relatively natural habitat just outside the city and a built-up neighborhood within the city. Such counts have been made on the NCAR mesa, an area of natural grassland and ponderosa pine occupied only by the NCAR building, and an area of housing in the city just north and east of the NCAR mesa along Kohler Drive and Vassar Drive.

The observed population at NCAR mesa, counted from August through December every year, approximately 50 times per year, has varied considerably, but the annual count averages have stayed within the range of 15 to 21 deer (Fig. 2). Some counts on individual days have been as high as 76 deer, but these have been averaged with other days when only 5 or 6 deer were seen. We do not suggest that these counts represent all the deer using NCAR mesa—rather, they are a relative index of counts made in 1 hour, usually between 4:00 and 6:00 p.m., when the deer are most likely to be seen in the NCAR meadows. The numbers of deer seen peaked in 1984 and dropped back in 1987 and 1988 to levels about the same as or lower than they were 10 years ago. Two sharp drops occurred in the average

NCAR sightings in 1983 and 1985; we do not have a satisfactory explanation for these declines.

In contrast to the NCAR sightings, the counts of deer along the Kohler Drive neighborhood have risen from an average of 3.3 in 1981–1983 to an average of 8.0 in 1988. Whereas the NCAR herd has shown no consistent growth trend, the urban deer along Kohler Drive have more than doubled (actually a 142 percent increase). These figures suggest that more deer are coming into the city, increasing the perception of rapid population growth.

Another type of evidence suggesting that the Boulder deer population has become more urbanized are data on deer roadkills on the streets of Boulder. This number has unfortunately risen dramatically in recent years. In 1985, only 80 deer were reported as roadkills on Boulder's streets; in 1988, this number skyrocketed to more than 200 (Fig. 3). These kills have occurred predominantly in winter months, although the figures for 1987 and 1988 show a trend toward double-digit roadkills monthly even in some summer months.

The Urbanization of Mule Deer

Thus several types of data suggest, although they do not prove, that part of Boulder's deer population increase is due to greater numbers of deer moving into the city. Studies on the movements and home ranges of tagged deer show more of them adapting to the urban environment. A detailed study of nine tagged deer in 1986 showed that only one deer of the nine spent more than 90 percent of its time in the city; the others spent most of their time in open space. In 1987, six of these nine deer had moved into the city and were spending 90 percent of their time there. Although this is a small sample, it shows the trend toward occupying the human environment. Similar results had been obtained in an earlier study in 1983 and 1984 by Brian Coppam and Diane Gentile in north Boulder, in which they found among eighteen tagged deer that some individuals consistently moved into inhabited areas, whereas others rarely did. For some deer, it seemed as if the city life was becoming a habit.

Why should this be so? What factors might prompt wild deer to move into the suburbs and city? Several possibilities come to mind, although existing research can neither prove nor disprove them.

First of all, the deer may be attracted by the greater variety of vegetation and potential food plants. Many homeowners know, to their dismay, the predeliction of mule deer for shrubs, young trees and many garden plants. City deer may have a nutritional advantage over country deer.

Secondly, houses provide shelter from chinook winds and severe winter weather. We have the impression that more deer are seen in the city during heavy snows, severe cold and periods of strong winds, but this point is not well researched. The possibility is reinforced, however, by Boulder City roadkill data, which show a distinct rise in roadkills in winter months (Fig. 4).

Thirdly, protection from hunting and predation may be a factor bringing deer into the city. There is certainly more hunting activity and mountain lion predation in the foothills than the suburbs.

Finally, it seems almost certain that habitat imprinting can occur for young born in the city. Adult animals of many species tend to prefer habitats in which they were born and raised, and if female deer enter the city to bear fawns, it is logical to surmise that these fawns grow up to be city deer.

The Condition of Boulder Deer

Despite the high population density and some conflicts with humans, the deer of Boulder seem to be in relatively good shape. Injured animals are seen, perhaps survivors of automobile accidents or individuals that have clashed with fences, but these are not conspicuously abundant. Injured animals may survive better in the city than in a wild environment, since natural predators are few. For example, for two years we have seen a doe with a completely incapacitated hind leg hobbling along and doing quite well in the western suburbs around Norton Avenue. Such an animal would not last very long in natural circumstances.

During the winter of 1988-89, bone marrow samples were taken from 40 roadkilled deer both inside and outside the city. (Solid white marrow indicates fat deposits and good nutritive condition; gelatinous red or yellow marrow indicates an undernourished condition.) On a scale from 1 (peak nutrition) to 5 (total fat depletion and poor nutrition), Boulder's deer averaged 2.34, with no clear differences between city and country animals. This indicates a reasonable state of nutrition—not ideal, but certainly not starvation.

A leading indicator of the health of a deer herd is productivity in terms of fawns. The August and September fawn-doe ratio of the NCAR herd has averaged 0.61 ±0.05 over the past 10 years, with a low of 0.37 and a high of 0.83. Most years the fall fawn-doe ratio was in the range of 0.7 to 0.8. This is an acceptable average for mule deer—lower than that of the best, well-fed herds in excellent habitat, but not suggestive of a starving herd. The years of lowest production (<0.50), which occurred between 1983 and 1985 and which suggested natural population regulation of the deer, followed severe winters, during which the deer feed more readily on ponderosa pine extending above the snow. In cattle and some other animals, the secondary compounds of ponderosa are known to cause miscarriages in early pregnancy. It is not known whether the same effect occurs in mule deer, but the physiological possibility exists.

Another concern about high-density deer populations is the possibility of forest damage due to overbrowsing. The "browse lines" of north-central and northeastern forests due to overpopulations of white-tailed deer were classic features of many forests in Wisconsin, Michigan, Pennsylvania, New York and New England, and they were often associated with serious starvation problems of deer some years ago.

In the Boulder open space environment there have been neither widespread browse line problems nor deer starvation problems, suggesting that our deer population has not reached advanced stages of overpopulation.

Management and the Future

The Boulder deer situation is not a cause for celebration or complacency. Any animal population in which automobile accidents cause up to 20 percent mortality is not desirable for either the wild animal or the human being. Numerous suggestions for solutions have been offered, from outright reduction of the deer herd to various devices to reduce automobile accidents. No system seems to provide the ideal solution, but several ideas may have merit.

A direct hunting program to reduce the herd is not feasible. Hunting cannot be done within the city nor in adjacent open space, and if it is allowed in more distant open space, the wrong deer will be shot. The most important deer to regulate or reduce are the chronic urban dwellers. Possibly they can be trapped and released far away, but this is expensive and most likely would be a slow death sentence to city-adapted deer in any case. Various devices, such as reflectors, to frighten deer from highways are being tested, but deer seem to become habituated to them, and their value is not yet proven. The best long-term solution may be to find more appropriate ways to discourage deer from coming into the suburbs—fencing in or using effective repellents around attractive gardens and shrubbery and certainly calling a halt to feeding the deer, as enjoyable as this may be. Table 2 lists plants whose attraction for deer seems to be minimal. Perhaps we need to find a satisfactory way to apply nontoxic birth control measures for city deer. This might involve trapping does and using long-term progesterone implants—again a fairly expensive proposition, but one that could be employed as a routine part of a trapping and marking program.

An important part of any wildlife management program is public education and cooperation, and the Boulder deer situation provides a clear example of this principle. Various possibilities, from driver education to landscape planning and gardening design, offer different options to reduce the unfavorable impacts of Boulder's deer population while at the same time retaining and even enhancing its desirable and enjoyable features.

ACKNOWLEDGMENTS

We are indebted to many students and colleagues who have assisted in these studies over the past ten years—at the University of Colorado, Karen Rasmussen, Steven Lipshur, Michael Sisk, Brian Coppam, Diane Gentile, Gail Fontaine, Rachel Kirby, Andrew Ames, Frances Schmeckel, Michael Knapp, Michael Mooring, Dorie Brownell, Thomas Ryan, Bob Palkowski and Kelly Peterson; among Boulder Mountain Parks rangers, Ann Wichmann, Dick Lyman, Jeanne Scholl, Mary McNellan, Eric Peterson, Ann Armstrong, Rick Day, Charles Duncan, Mat Claussen and Pete Perry.

Open Space staff who have assisted are Janet George, Mark Reddingen, Scott Wait, Rich Smith, Jack Kissell, Mark Gershman, Brent Wheeler, Bill Grabow, Bill Dimond and Jeff Holland.

We are also grateful for the cooperation of Ron Green and Western Resource Development Corporation of Boulder, who played an important role in the first trapping and tagging studies in 1983 and 1984. Finally, we are indebted to the Colorado Division of Wildlife staff who assisted with trapping and who generously supplied radiotelemetry equipment for studies of movements and home ranges.

WENDY SHATTIL/BOB ROZINSKI

A buck mule deer in summer velvet trots across a road to reach another part of its range.

TABLE 1. Some Comparisons of Mule Deer and White-tailed Deer

	Mule Deer	White-tailed Deer
Scientific name	*Odocoileus hemionus*	*Odocoileus virginianus*
Height at shoulders (adult males)	3 feet	3 feet
Body weight (adult males)	up to 300 lbs.	up to 300 lbs.
Body weight (adult females)	100-150 lbs.	80-140 lbs.
Ear length	8-10 in.	4-6 in.
Tail	narrow white with black tip	broad brown—dorsal white—ventral
Antlers in male	dichotomous, Y-shaped parts	several tines from a single main beam
Behavior	more gregarious, larger social groups than white-tailed deer	rarely seen in Boulder in groups of more than 5 or 6; often fewer
	very commensal with people; often enter housing areas	shy and wild; not seen in city

TABLE 2. Deer-resistant Plants*

Moderate water demand (.75 in./week in hot weather)
Potentilla (*Potentilla fruticosa*)
Lilac (*Syringa vulgaris*)
Concolor fir (*Abies concolor*)
Colorado spruce (*Picea pungens*)
Creeping mahonia (*Mahonia repens*)
Virginia creeper (*Parthenocissus quinquefolia*)
Peony (*Paeonia* spp.)
Winged euonymus (*Euonymus alatus*)
Rose-of-sharon (*Hibiscus syriacus*)
Van Houtte spirea (*Spiraea 'Van Houttei'*)
Russian olive (*Elaeagnus angustifolia*)
Lavender (*Lavandula* spp.)
Santolina (*Santolina* spp.)
Russian sage (*Perovskia atriplicifolia*)
Anthony Waterer spirea (*Spiraea 'Anthony Waterer'*)
Low water demand (0 to .5 in./week in hot weather)
Bluemist spirea (*Caryopteris* sp.)
Rabbitbrush (*Chrysothamnus* spp.)
Piñon pine (*Pinus edulis*)
Curl-leaf mountain mahogany (*Cercocarpus ledifolius*)
Yarrows (*Achillea* spp.)
Iris (*Iris* spp.)
Daffodils (*Narcissus* spp.)
Snow-in-summer (*Cerastium tomentosum*)
Chocolate flower (*Berlandiera lyrata*)
Prairie zinnia (*Zinnia grandiflora*)

*Plants that are rarely damaged in Boulder, Colorado. Plants that are fertilized and watered heavily are much more likely to be eaten than those that are grown under natural conditions.

Plants are shown in their zone of general preference, but many of these plants will thrive in several zones. Table prepared by Jim Knopf.

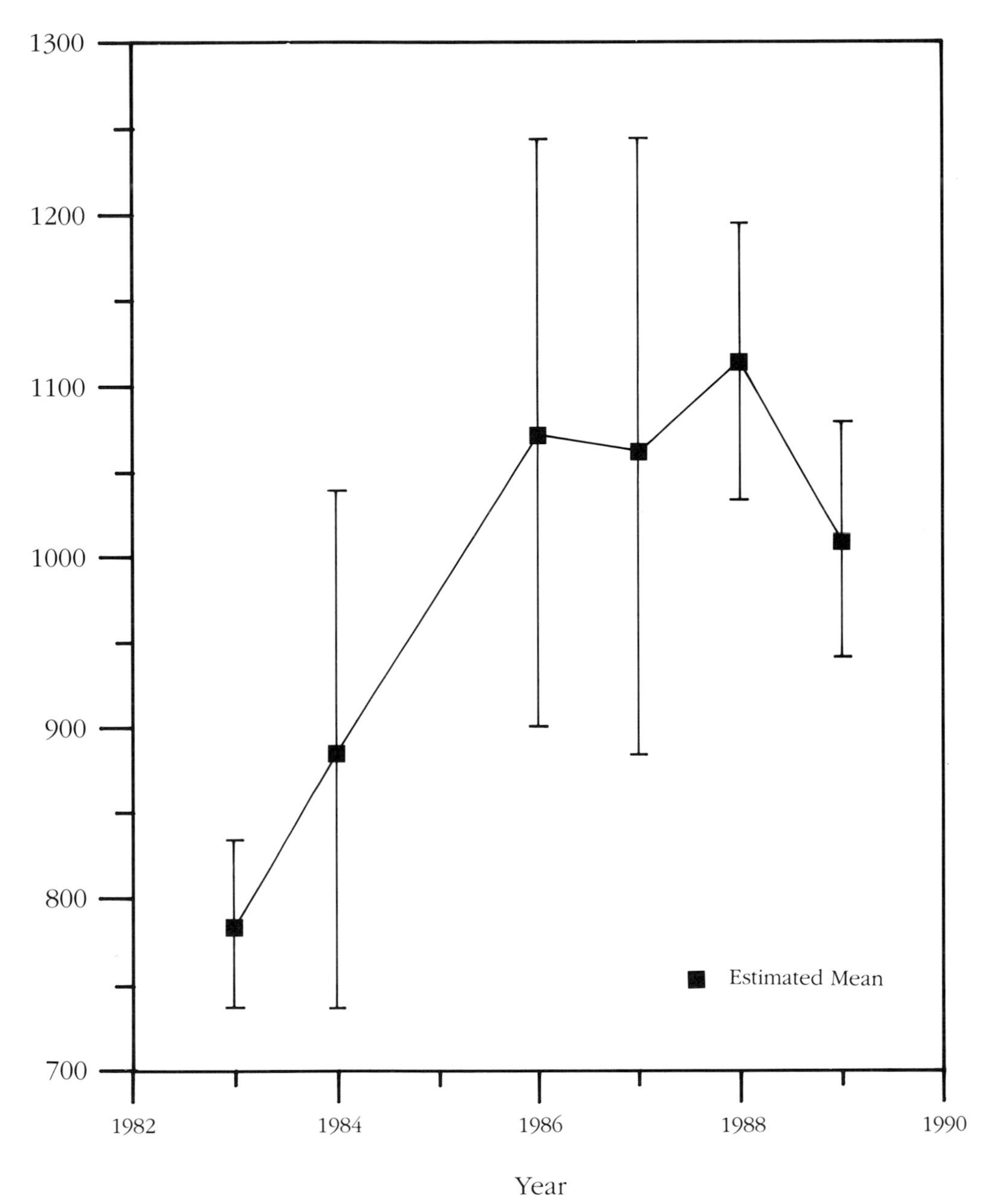

FIGURE 1. Estimated Mean or Average Population of Boulder Mule Deer, 1982-1989. Vertical lines show standard error of mean.

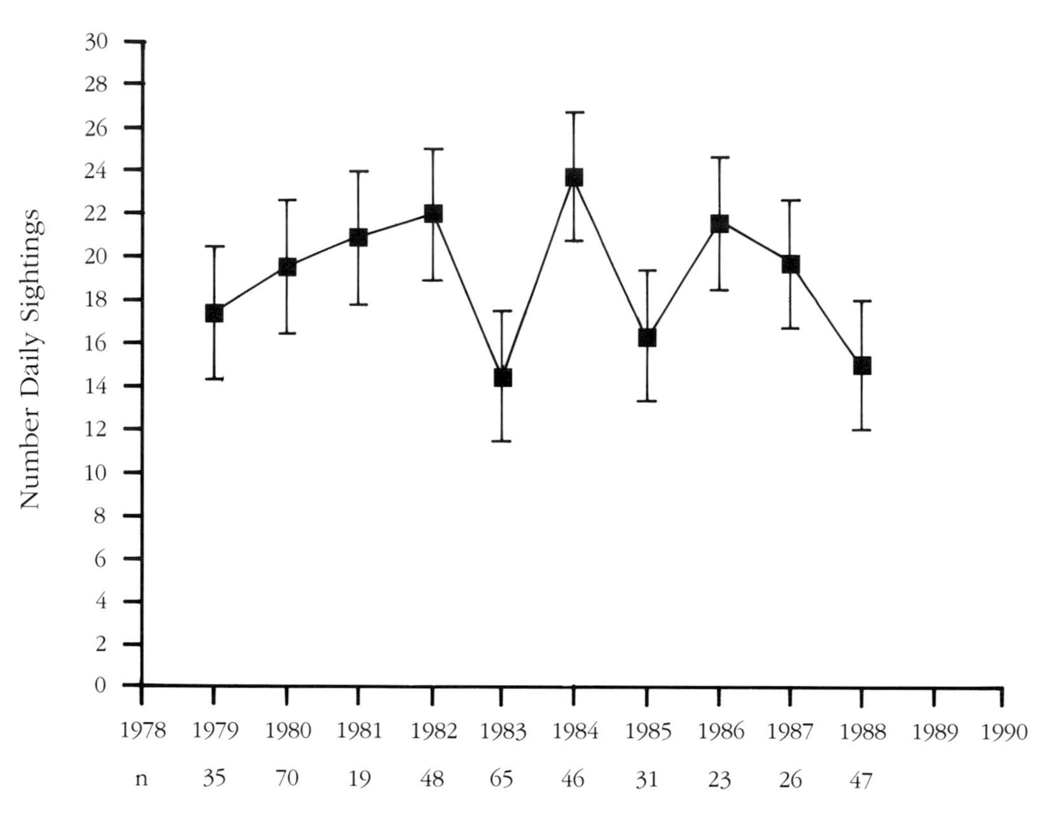

FIGURE 2. Average Number of Daily Sightings of NCAR Deer Herd, August-December counts, 1979-88. Vertical lines show standard error of mean.

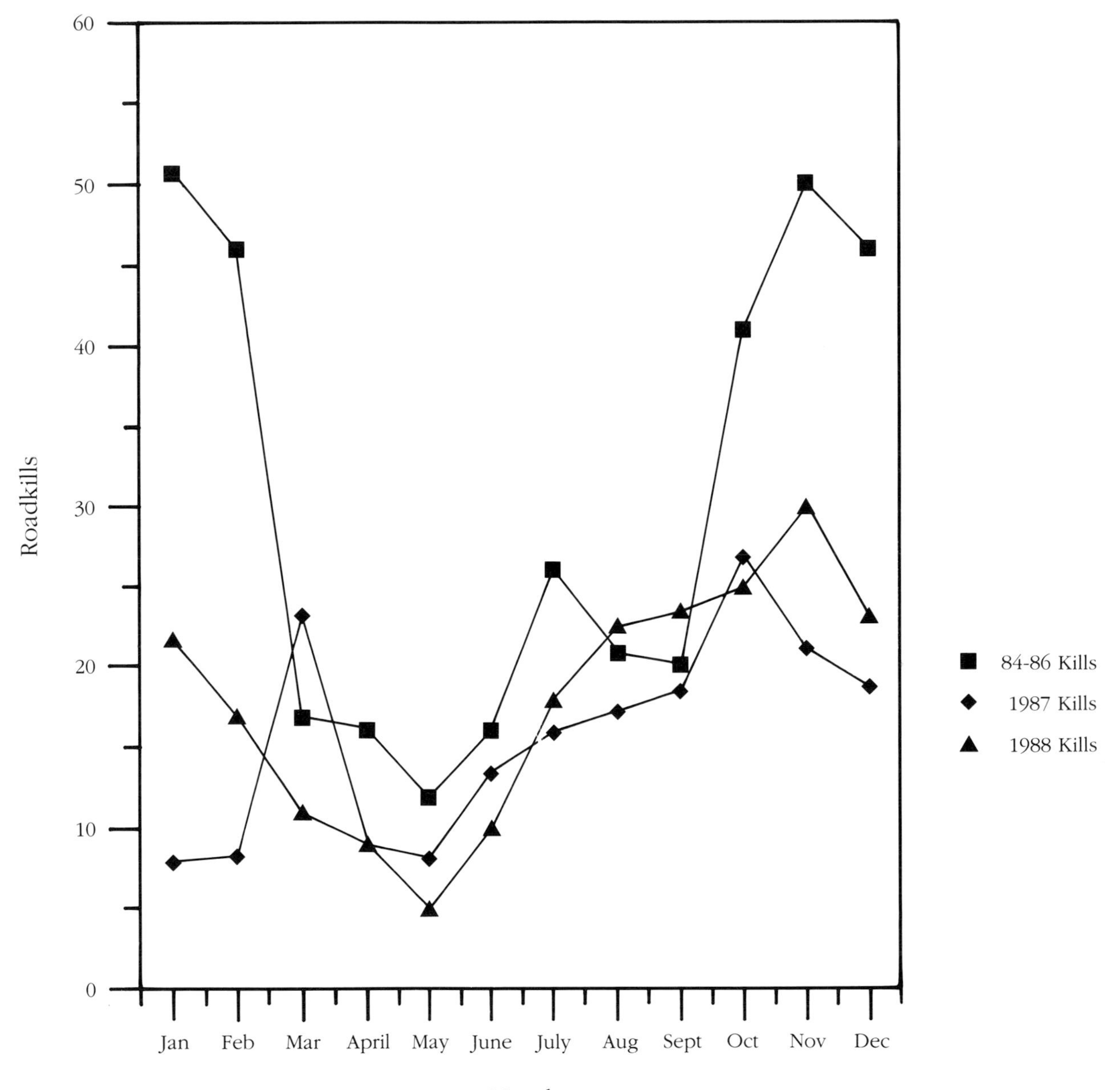

FIGURE 3. Reported Mule Deer Roadkills in Boulder, 1983-88.

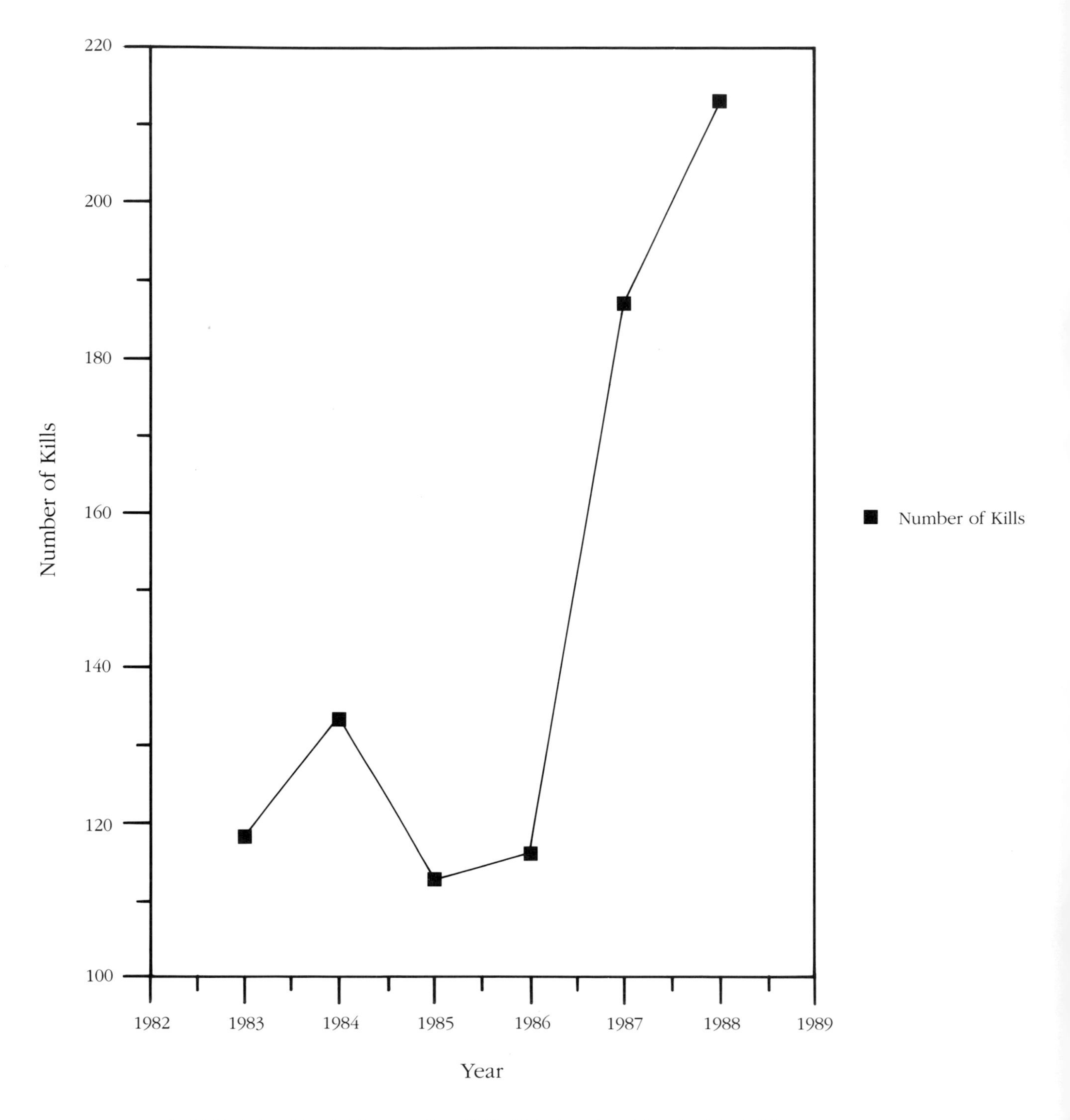

FIGURE 4. Boulder Mule Deer Roadkills Tabulated by Month, 1984-88.

NINE

Peregrine Recovery IN DOWNTOWN DENVER

JERRY CRAIG

When the Colorado Wildlife Federation first proposed the idea of releasing peregrine falcons into downtown Denver, responses from the Colorado Division of Wildlife staff (including me) ranged from gasps of horror to a muttered "no way." After all, the peregrines we're used to seeing inhabit glorious settings such as the Royal Gorge, Black Canyon of the Gunnison National Monument, Colorado National Monument and Dinosaur National Monument. Releasing them in the concrete-and-steel canyons of Denver seemed almost a crime.

The suggestion came from Wayne Sandfort, a board member of the Colorado Wildlife Federation, who hatched the idea after visiting the nest site of a wild pair of peregrines on the roof of the Hotel Utah in downtown Salt Lake City. Fortunately, the federation persisted in spite of the Division of Wildlife's initial reactions. The downtown peregrine project has turned out to be one of the most popular and educational efforts undertaken for wildlife in Colorado.

The proposal was to "hack" nestling peregrines into the city. Hacking is used when the young do not have the benefit of protection or care from adult birds. Falcons about 35 days old are placed in an enclosure on a suitable ledge at the reintroduction site. They are fed and cared for by humans until they have learned to fly and are capable of fending for themselves. This procedure assures that the young become familiar with their surroundings and will have a tendency to return as adults.

In the eastern United States, where the idea of hacking peregrines first came up, great horned owls thwarted many efforts to release the falcons back to historical cliff nest sites. Without benefit of adults for protection, the fledglings were easy prey for the owls at night. Relocating the owls was ineffective, since their vacant territories were quickly reoccupied. Biologists soon discovered, however, that falcons could be successfully released in cities and even from towers in coastal salt marshes. Since falcons have a propensity to return to structures similar to those from which they have been released, city-dwelling peregrines became almost commonplace. As the eastern population expanded, falcons returned to historical breeding cliffs as adults that could defend themselves against owls.

Although we tend to associate them with wilderness areas, peregrines are no strangers to urban settings. Castles and towers in England, Spain and Germany have been inhabited by breeding falcons since medieval times. The birds also have shown a predeliction for modern skyscrapers. The Montreal headquarters building of the Sun Life Assurance Company of Canada was occupied from 1939 to 1955 by a female that reared 21 young over that period. Peregrines bred on city hall in Philadelphia in 1946, and Mormon Temple in Salt Lake City hosted young peregrines in the 1960s. Although they were not documented as breeding in Denver, peregrines were occasional winter visitors to the city before their population declined.

Reproductive failure brought on by pesticide-induced eggshell thinning caused peregrine populations throughout North America to crash, and the species was declared endangered in 1972. Colorado's recovery efforts began that same year. Nevertheless, the state's peregrine population had declined to only four breeding pairs by 1979, down from an estimated 80 pairs. A project begun in 1978 involved removing thin-shelled eggs from nests, substituting plastic replicas to occupy the nesting parents, artificially incubating and hatching the eggs, and returning the young to the wild pairs. Attempts were also begun to reestablish falcons at vacant nest sites through the technique of hacking.

The proposal to release peregrines in Denver came after a decade and a half of recovery efforts in Colorado that had attained some success. By 1988, the Western Slope population had grown to 22 breeding pairs of falcons. However, the population on the Eastern Slope had continued to decline, and all pairs had disappeared by 1981. With the Western Slope out of jeopardy, recovery actions shifted to the Eastern Slope population.

Proponents of the Denver release suggested that it could augment the recovery effort along the Front Range and generate additional funds from private donations to support it. The Division of Wildlife remained skeptical, but a visit to the proposed release site on top of the 23-story building at 1560 Broadway was encouraging. I found the surroundings remarkably similar to canyon country. The adjacent office towers looked very much like cliff faces. At that height, people on the sidewalk were mere specks 230 feet below, and traffic noise was muted. I was also surprised by the amount of potential hunting habitat offered by Denver's parks and greenbelts.

I came away supporting such an endeavor, but with a few concerns and questions: (1) Given the need to maximize efforts to restore the Eastern Slope population to vacant historical sites, the Denver release had to be a relatively low priority. (2) Would urban-released falcons or their offspring contribute to the recovery program and relocate to vacant wild nest sites on the Front Range adjacent to Denver or would they only occupy urban settings? (3) Would Denver pigeon fanciers take matters into their own hands as they did in Los Angeles and begin shooting peregrines when they occasionally captured birds from their domestic flocks? (4) Would peregrines suffer effects of poisons used to control wild pigeon populations? (5) Given the hazards of the city setting, would urban-released falcons suffer higher mortality rates than wild-released falcons ? I felt that if these concerns could be put to rest, an urban release could have merit as part of the statewide recovery program. Still, I worried that the publicity could lead the public to believe peregrines reside only in cities and the big picture would be lost.

Discussions with biologists experienced with urban peregrine releases in other states gave us cause for optimism about the undertaking. In Los Angeles, up to six pairs of peregrines bred in the city at one time. The Los Angeles birds were composed almost equally of wild-hacked birds and urban-released falcons. Efforts in New York, Boston, Baltimore, Chicago and other cities yielded survival rates approximating 80 percent, which is similar to wild-release efforts. Only the causes of mortality differed. Instead of being caught by great horned owls and golden eagles, city-dwelling falcons had to contend with potentially fatal hazards like mirrored windows, ventilator shafts, traffic and irate, gun-brandishing pigeon fanciers.

Pigeon fanciers appeared not to be a problem here, because the aficionados in Denver concentrate upon homing pigeons and not the "tumbler"-type pigeons that are popular in Los Angeles. Those who race homing pigeons usually transport their flocks a great distance from their lofts before releasing them; this means the birds' exposure to peregrine attack is minimal. Tumbler pigeons, on the other hand, are released from their lofts to fly about the immediate vicinity and exercise. They have poor homing instincts and tend to stay close to their lofts, attracting resident peregrines. A peregrine's attack will send the pigeons into a panic and the entire flock may fail to return to the loft.

One misconception that arose was that the peregrines would control the wild urban pigeon population. At best, a peregrine will kill and consume one pigeon a day, a loss the wild pigeon population can afford. The peregrines probably do teach the pigeons to be more cautious when crossing open air space. By and large, urban peregrines feed on smaller birds such as starlings, swallows, jays, robins and warblers. Concern that peregrines would suffer effects of the poisons used to control urban pigeon populations was allayed by the fact that most control actions are taken in the winter, when peregrines likely would not be present. Avitrol, the preferred substance in pigeon control, is not too harmful to predators that take advantage of sick or woozy pigeons. As a precaution, letters were sent to all who were likely to be involved with pigeon control to alert them to the peregrine release and encourage their support. To date, poisoning has not been a problem.

Once the project's feasibility was established, a partnership was formed between the Colorado Wildlife Federation, Denver Museum of Natural History, and Colorado Division of Wildlife. The Peregrine Partnership served as a vehicle to receive private donations and facilitate the project. Members of each organization combined their skills in educational programming, marketing, fund-raising, public relations and biology to make the program a success. The partnership facilitated numerous media events, a fund-raising reception at the governor's mansion, an exhibit and reception at the Denver Museum of Natural History and a peregrine art show with proceeds going to the release effort. The Museum of Natural History developed a peregrine exhibit detailing natural history, eggshell thinning and the Denver release project. Curriculum specialists from the Denver Public Schools worked with the educational staff from the museum and the Division of Wildlife to develop educational packets for distribution to primary and secondary school teachers.

Hack boxes in wild settings are usually uncomplicated affairs approximately 4 feet wide, 4 feet high and 6 feet long. The Denver hack was something else! It had to accommodate the large access door to the roof, and mirrored windows were required to assure that the birds could not see and be frightened by observers. An architect eventually was engaged to design the box; building and zoning waivers had to be obtained. The final product resembled a penthouse more than a box.

A number of appearances on radio talk shows and on television news permitted the Peregrine Partnership and the Division of Wildlife to prepare the public for the releases and respond to questions and concerns. Among other things, we allayed fears of several listeners that a falcon would swoop down and snatch up their chihuahuas or poodles; we explained that peregrines feed exclusively on medium- to small-size birds that they capture in flight.

With the project moving ahead on all fronts, a relatively serious problem still had to be overcome. Young birds born in captivity were in short supply because the national recovery focus had shifted to the northern states. The two dozen falcons that were available to Colorado were all slated for wild sites. Since Colorado's primary objective is to reestablish peregrines at vacant historical nest sites, the Denver release was a lower priority. An alternate source of falcons had to be developed. The National Park Service came to the rescue and offered to permit us to "recycle" several wild breeding pairs of falcons in Dinosaur National Monument to provide the falcons needed for Denver. We visited nesting sites after the pairs had laid their eggs. All the eggs were removed and flown to Boise, Idaho—headquarters of the Peregrine Fund's captive propagation program—to be hatched in captivity. After two weeks, the pairs "recycled" and laid second clutches of eggs. Captive-hatched chicks were returned to some pairs and and their second clutches were removed and hatched in captivity. Other pairs were permitted to hatch and rear their second clutches of eggs. Such recycling efforts provide surplus young for release elsewhere without impacting populations.

The arrival of the young falcons from Boise was a media event in both 1988 and 1989. They were welcomed by the media, the governor and the mayor both years, and even by some Soviet visitors in 1989. The falcons were banded and confined to the hack box so they would acclimate to their surroundings. Television station KCNC put cameras at the hack box and transmitted live videos of the birds to monitors on the ground floor and at the Museum of Natural History's peregrine exhibit. After four or five days, the falcons were equipped with radio transmitters and the barred front of the box was removed to allow access to the roof. Invariably, the youngsters moved to the roof edge for a better view. Then they would return to the vicinity of the box to bathe and to eat catered meals of quail. They were provided with quail for up to six weeks or until they were successfully hunting for themselves.

A cadre of volunteers had been established to keep track of the falcons on their sojourns throughout the city, rescue them if they landed in the street and answer questions from the public. A very dedicated group soon emerged that was on hand almost every day. One couple would work at the site all day, go home for dinner and then return to stay until dark. Although they weren't paid, they were rewarded with the sight of many impressive flights.

Radio communications were imperative so that hack site supervisors at the command center could maintain contact with the volunteers. First-class radios were loaned by a firm during the release and a cellular telephone was also loaned to call for help in emergencies. Observation posts were provided on a vacant floor of the Colorado State Bank building. (When that space became unavailable in 1989, the management of the ARCO building provided office space that overlooked the hack site.)

In the first year, some of the early flights were spectacular; the falcons would fly out to circle the state capitol, then return to the roof. Other flights were not as picturesque. One male was blown over backward by a gust of wind off the roof of 1560 Broadway and ended up on a window ledge several stories below. He was flushed from his perch the next morning and was finally recaptured and returned to the roof five hours later, but not before the Denver fire and police departments, as well as 30 volunteers, had gotten involved. Another female was rescued from a garbage dumpster after a rather inglorious first landing.

Each of the three females and two males that were released in 1988 got into trouble and had to be rescued at least once, but thanks to transmitters and dedicated volunteers, each achieved independence. By early fall, all five had dispersed. One male was observed in midwinter several times and likely wintered in the area, while his siblings probably migrated to warmer climes.

Five falcons from the recycling efforts in Dinosaur National Monument were again placed in the hack box in 1989. This contingent consisted of four males and one female. Things did not go as well as they had the previous year. Two birds made it to independence, two disappeared (presumably they either flew out of the area or were killed) and one bird died. The highlight of the 1989 effort was the appearance downtown of a wild yearling male who quickly accepted the urban youngsters and flew with them almost daily.

The single greatest impediment to the 1988 and 1989 releases, and I suspect to future attempts, was the great number of mirrored windows that are in vogue on Denver office buildings. When viewed from certain angles, they show only a reflection of the sky, inviting inexperienced falcons to fly through them. All the birds found out the hard way that they could not fly through glass. Most suffered only glancing blows without apparent ill effects, but three falcons struck windows with sufficient force to require rehabilitation. Two suffered impaired vision and broken mandibles (beaks), and one sustained a fractured leg. After convalescing at the Birds of Prey Rehabilitation Foundation in Broomfield, all were released back to the city.

As predators, peregrines face fairly high mortality rates. Approximately 60 percent of the falcons die within their first year. Before they realize the full potential of their magnificent powers of flight, they are vulnerable to attack by golden eagles and may be killed by great horned owls if they choose an exposed night roost. The most common cause of death, however, is collisions with objects while in flight. This is easy to understand when you consider that peregrines can fly up to 60 miles an hour in level flight and may achieve speeds in excess of 180 miles an hour in a hunting dive, termed a stoop. Objects as small as twigs can shatter wings at such speeds.

In order to compensate for the high mortality, falcons will have to be released in the city for at least three years before a bird of breeding age can be expected to select Denver as a nest area. All that is needed is one adult taking up residence, because that bird will then maintain a territory and try to attract other peregrines that pass through the region. Should he survive, the yearling that was present in 1989 may return as an adult in 1990 and pair with one of the females that was released in 1988. Only time will tell. Although the falcons have a propensity to return to areas from which they fledge, the failure of birds to return does not necessarily mean that they have died. Peregrines do a lot of wandering in their first two years, and when mature, they may take up residence with lone adults elsewhere. Banding efforts have revealed that the falcons may winter at some localities within the state, such as the San Luis Valley and the Grand Junction area, but the majority move southward. One youngster's first southward migration took him as far as Panama, but it appears that most may winter in north-central Mexico.

In addition to augmenting wild releases, it is possible that urban-released falcons may avoid some of the pesticide contamination that impacts their eggshell thickness. Falcons that are reared in an urban environment likely will seek such an environment even if they winter away from Denver. Since the majority of the prey birds that inhabit metropolitan areas are nonmigratory, they are likely not to harbor the high body burdens of the chemicals that migrants encounter. This idea can only be tested when peregrines take up residence in Denver and the condition of their eggshells can be monitored.

Another positive aspect of the Denver project is that it allows this endangered species to be seen by more people. The mountain sites that peregrines normally occupy are relatively uninhabited, and if one of the falcons is seen, the sighting is usually only a fleeting glimpse of a fast-moving bird in flight. Urban-dwelling falcons increase the opportunity for people to observe and cherish peregrines. Seeing falcons diving and soaring over Denver should be some compensation to taxpayers for their long and unwavering support of the nongame program and endangered species recovery funded through the state income tax checkoff.

In the spring of 1989, the U.S. Forest Service, Division of Wildlife, U.S. Air Force Academy and Aiken Audubon Society applied experience from the Denver project and hacked peregrines at a vacant historical site on a mountain cliff immediately west of the academy's visitor center. An exhibit has been installed at the visitor center, and the public can see wild falcons nearly at the front steps. The project is supported through private donations and has the added benefit of reestablishing the species directly into a vacant historical site.

Compared with similar species such as prairie falcons, peregrines never were abundant in Colorado. As breeding birds they were restricted to the foothills and mountains of the Eastern Slope and the mountains and canyons of the Western Slope. Even prior to their decline, only an estimated 80 pairs resided in the state.

Given the success we've had on the Western Slope, if we can reestablish 12 successfully breeding pairs along the Front Range, we can consider our recovery efforts a success.

We must remember that the release activities in the wild and in Denver are an endeavor to reestablish sufficient numbers of peregrines at their favored breeding sites along the Eastern Slope. The project is only one step in the recovery process. We must still address the ultimate cause for the species' endangered status: pesticide-induced reproductive difficulties. That problem is international in scope and will not be solved in Colorado. Thus, the teacher packets developed by the Peregrine Partnership may have a more profound impact than the actual recovery actions, for they will educate our children about the interrelationship of biosystems and the global implications of man's activities.

WENDY SHATTIL/BOB ROZINSKI

PHOTOGRAPH LEGENDS

PAGE 125

A squadron of Canada geese prepares for touchdown at the Adams County Golf Course during subzero weather. Even bitter cold will not drive Denver's resident geese south as long as they have open water on lakes and plenty of bluegrass to eat on lawns, parks and golf courses.

PAGE 126

Chatfield State Recreation Area is multiple use: fishing, boating, water skiing, camping, picnicking, hiking, biking, horseback riding, model-airplane flying and hot-air ballooning. The wealth of tolerant wildlife there is heartening, especially in the great blue heron rookery, surrounded as it is by summertime frenzy.

PAGE 127

Canada geese are everywhere, so it seems, even in ornamental reflecting pools of ultramodern office parks, such as here in Denver's Greenwood Plaza. One goose felt secure enough to nest in this place, which is a tribute to the sensitivity of daily passers-by.

PAGE 128 ABOVE

A picnic in the backyard usually means insect guests. These libidinous ladybugs landed on the photographer's plate, but not to lunch. When they do dine, ladybugs eat huge numbers of harmful insects; they are a valuable natural pesticide.

PAGE 128 BELOW

This caterpillar was caught devouring the photographer's ash tree leaves. It was turned loose after posing on an old thermometer. A few leaves are a small price to pay for such beauty.

PAGE 129 ABOVE

A gorgeous swallowtail butterfly sips nectar from a day lily in a Boulder garden. Butterflies are second only to bees as pollinators. Gardeners can selectively plant flowers and shrubs that butterflies find irresistible to create butterfly gardens.

PAGE 129 BELOW

A large green katydid blends perfectly with the leaf of a rose bush in a suburban Denver garden. Only the males sing, and only at night. In a summer, a male may scrape its wings together 30 to 50 million times to produce its familiar song.

PAGE 130 ABOVE

A painted turtle slips into the water after its nap on a giant water lily pad at the Denver Botanic Gardens. A garden staffer suggested that this is a dimestore pet turtle released by its owners. However, the painted turtle is a common native species.

PAGE 130 BELOW

A rare and beautiful migrant to the Denver area is this scissor-tailed flycatcher at Chatfield State Recreation Area. One was seen there in 1988 and 1989. Spotting such a rarity outside its normal range is a big thrill for birders.

PAGE 131 ABOVE

Nearly every lake, pond and creek in the Denver area harbors muskrats, which are overgrown, amphibious field mice. They may dig a burrow in the bank, build a lodge or, like this one, live in a drainage culvert. They are smaller than beavers and have a long ratlike tail.

PAGE 131 BELOW

Like a child eating spaghetti, a large carp slurps in a cattail leaf. Carp, along with ducks, cormorants and muskrats, inhabit a pond at the Wellshire Golf Course. Note the floating golf balls.

PAGE 132 ABOVE

This broad-tailed hummingbird built her nest on the end of an electrical wire hanging from the porch ceiling of an abandoned ranch house near Roxborough State Park. The lichen-covered nest is only about 3 inches in circumference. Her babies are safe here from virtually any four-legged predator.

PAGE 132 BELOW

Tracks of skunk, chipmunk and child mark a trail at Roxborough State Park. Tracks give city dwellers a way to learn about their many seldom seen neighbors. We share pathways more often than we know.

WAYNE AIGAKI

SHERM SPOELSTRA

GARY A. HAINES

CECILIA T. ARMBRUST

BINI ABBOTT

JIM KNOPF

DAVID H. SAKAGUCHI

J. B. HAYES

BILL BEVINGTON

TINA JONES

J. B. HAYES

BILL BEVINGTON

CECILIA T. AMBRUST

T E N

Of Hope And Irony:
ROCKY MOUNTAIN ARSENAL

GARY GERHARDT

Every nose was pressed against the window glass of the bus; every passenger strained to witness the drama being played out a few feet off the roadway. Two ferruginous hawks were battling over the carcass of a prairie dog.

As the tour group watched, one of those rare moments of which birders dream actually happened: A magnificent bald eagle suddenly floated down between the hawks and laid claim to the disputed prize. Wings spread, feathers arched over its large golden eyes, the eagle issued a few short cries of warning to the smaller raptors. But before the eagle could begin to dine, still another player intruded—a coyote scrambled in among the birds, snatched up the prairie dog and trotted off, glancing over its shoulder at the cheated raptors.

"It all happened so fast, and right before our eyes," declared a Denver Audubon Society member who was on the tour. "I don't know how long you'd have to wait to see something like that in the wild again." Normally, perhaps a lifetime. But not where this bus had stopped. It was on a road running through 5,000 acres of "dog towns" on the Rocky Mountain Arsenal, 7 miles northeast of Denver's smoked-glass high-rises. Such interactions are a common occurrence there. More than 200 species of vertebrates live within the arsenal's 17,000 acres, which include what a former arsenal commander termed the most contaminated piece of ground in this country, but which others see as a "mini-Yellowstone" in the midst of 1.6 million people.

For 40 years, the U.S. Army produced deadly mustard and nerve gas bombs there, and Shell Oil Company manufactured pesticides and insecticides so lethal they've now been banned by the Environmental Protection Agency. Millions of gallons of toxic wastes were dumped into holding ponds that became witches' caldrons, contaminating about one-fourth of the arsenal's 27 square miles. (Today, the ponds have been dried up and most of the pollution is in the substrata.)

The presence of an abundance of wildlife on such hideously damaged land is explained by the fact that the deadly wastes are limited to about two dozen distinct areas, and most of the animals live in the uncontaminated buffer zones around them. Not surprisingly, the pollution has killed animals that came into direct contact with it—sometimes in massive numbers. For example, on just two days in May 1975, the Fish and Wildlife Service found 291 carcasses of birds along the shorelines of the notorious Basin F. But random samples indicate creatures that don't venture into polluted zones—and that don't feed on animals from polluted areas—are relatively free from contamination. Of 14 mule deer biopsied in 1987, only one had signs of contamination. It was an older buck that frequented the Shell plant site and may have licked one of the aboveground pipes that contained pesticide residues and had corroded.

Some predators do eat prairie dogs and other prey that roam the contaminated areas, for coyotes found dead on the grounds test high for pesticides. Yet prairie dogs sampled away from polluted zones showed no signs of contamination at all. There have been problems of eggshell thinning in kestrels nesting on the grounds; on the other hand, blood samples from captured bald eagles haven't indicated any levels of contamination. Perhaps most telling, fleas carrying plague invaded the arsenal in 1988, killing about 70 percent of the prairie dogs. Some feel the presence of fleas on property contaminated by insecticides furnished adequate proof that the pollution is strictly limited to specified areas.

Paradoxically, it was the very production of lethal chemicals that was indirectly responsible for the influx of wildlife at the arsenal. Because of the deadly nature of the product, public access was severely limited, and wildlife in the area never learned to fear humans. Additionally, the lands not used began reverting to their natural state, and without domestic livestock to compete for the prairie grasslands, wildlife flourished.

Given the arsenal's history, however, it's little wonder first-time visitors arrive expecting to see the likes of two-headed deer. Instead, there are healthy-looking bucks supporting trophy-size racks. Slick-coated coyotes can be seen trotting along a rise or playing tug-of-war with prey, 50 pheasants have been spotted feeding in a single field, and a dozen hawks have been observed vying for a kill while magpies waited for the leftovers.

Because there was no reason to control prairie dogs at the arsenal, the population grew to more than 40,000 animals. They, in turn, attracted coyotes, foxes, badgers and weasels. More than 200 ferruginous hawks have been counted, along with about 100 wintering bald eagles (25 or more of which may be on the site at any one time); red-tailed, rough-legged and Swainson's hawks; kestrels; burrowing and great horned owls; and golden eagles. "A lot of juvenal bald eagles use the arsenal as their winter roosting area, and with the loss of wintering habitat throughout the lower 48 states, the arsenal has become a very important site for them," said Pete Gober, a U.S. Fish and Wildlife Service biologist who manages the wildlife at the arsenal in cooperation with the army.

The bald eagles are the chief attraction for many of the visitors to the government installation, who are often unprepared for the profusion and variety of other wildlife they find there. Hugh Kingery of the Denver Audubon Society has a list of 145 bird species that have been seen at the arsenal. He says he thinks an accurate census would turn up more than 200 bird species on the land. Kingery was surprised to find some species, such as the black-throated blue warbler, which is rare in Colorado. He said he suspected the lark bunting, the state's official bird, may be nesting there.

On the northern prairie fields and wetlands of the arsenal's southern tier, visitors often spot black-tailed jackrabbits and desert cottontails bounding among abandoned factory buildings, and scores of waterfowl, including white pelicans, floating on three small natural lakes. Along the shores, great blue heron spear fish while killdeers and western sandpipers scurry after aquatic insects. Drainage from the lakes has created a marshland lined with willows and cottonwoods, and the calls of thousands of red-winged blackbirds, western meadowlarks and a multitude of other songbirds float up from the cattails and sedgy meadows where a herd of more than 200 mule deer quietly graze. Less evident, but also present in abundance in the wetlands, are mice and voles, amphibians and snakes.

Jan Justice, a member of the Denver Field Ornithologists, was reminded of Africa on her first visit to the arsenal. "I went on safari to Kenya with a naturalist group in 1981, and the prairie lands and all the wildlife there reminds me very much of what we have here," she said.

Curt Simmons, a Denver graphic designer and amateur naturalist, made the "grand tour" with Pete Gober. "I was impressed with the healthy nature of the ecosystem more than with the pure numbers of wildlife," Simmons said. "Like everyone else, I had a preconceived idea it was a place where there should be signs with a skull and crossbones. Sort of like the La Brea tar pits." His assumptions were quickly reversed. "I have spent quite a bit of time outdoors," Simmons said, "but I saw more wildlife in an hour at the arsenal than I've seen in months elsewhere. It's so open out there on the prairie. I sat in one spot and watched a red-tailed hawk go down on a mouse, and a short distance away, a family of burrowing owls were standing outside a prairie dog hole watching me watch them. They showed no fear at all."

Gober believes the remnant habitats found at the arsenal afford a unique opportunity for people to see the serengeti that existed before it was cultivated and overgrazed. Whether this state of affairs will last, however, hinges on a number of factors, including how thoroughly the land can be decontaminated. Although production ended in 1982 and cleanup operations—predicted to cost $5 billion and take more than a decade to complete—are beginning, the future of both the land and the wildlife is still undecided.

According to Col. Daniel Voss, commander of the cleanup project, the arsenal will remain under army ownership until after cleanup is completed. And as long as the arsenal is owned by the army, the Fish and Wildlife Service will be present. After that, the status of the acreage is uncertain. Already, land use planners in the communities surrounding the arsenal are setting their agendas for development of the area, including railroads and highways running through the property.

Besides cleanup operations, Col. Voss is charged with wildlife management and increasing public access to and use of the arsenal land. He believes he owes it to the public to give accurate and adequate information so a sound decision can be made on the ultimate use of the facility. He's attempting that by discussing the issues and concerns about the cleanup at public meetings and in talks with community leaders. His efforts also include sponsoring school kids' tours and getting volunteers involved. "A lot of the thrust is in the wildlife arena because of the unique opportunities we have," he said. Col. Voss concedes there's a bias among the public against the arsenal "because, in the past, the arsenal has been a bad actor. We did some things that were not smart out here and we tried to cover it up or not address it. As a result, there are people who think we're still manufacturing chemical munitions," he said. "So I'm opening it up for people, inviting anyone to come and look."

One of the first decisions Col. Voss made when he came to the arsenal was to refurbish the old officers' club as a visitors' center. The facility, on C Street just south of 7th Avenue, offers a series of wildlife displays and a 23-minute video describing the wildlife as well as the history of the arsenal and the cleanup operations. It's located close to the three natural lakes, which feature catch-and-release fishing.

The second decision Col. Voss made was to build an observation area just east of the bald eagle roost along First Creek off Buckley Road. The viewing area is on a high point so that visitors can see almost all the way to the northern and southern ends of the arsenal, especially in the winter when the leaves are off the trees. It affords a view not only of the eagle roost but of any wildlife traveling or feeding along the First Creek drainage area. The viewing port is a belowground "bunker," contoured into the ground with grass and shrubs on top, to keep the eagles from spooking at the sight of a human silhouette. Because the viewing area is away from polluted areas and can be reached from Buckley Road outside the arsenal, it affords an opportunity for people to come to the arsenal without making prior arrangements.

Col. Voss speculates that 30 years from now people who have visited the arsenal may convince elected officials that the best use of the land is as open space and a wildlife refuge. They will surely be supported by the environmental groups, including the Colorado Wildlife Federation, Denver Field Ornithologists and Denver Audubon Society, that consider the arsenal an extraordinary urban wildlife treasure. Their feelings are perhaps best summed up by Jan Justice of the Denver Field Ornithologists: “It's essential we pick up whatever open space we can while we can. If there's another oil embargo, the petroleum industry will be back, building will boom, and people will be looking at the arsenal as just another piece of prime real estate to be developed. What we now have is a window of opportunity to save this property as an urban wildlife sanctuary.”

SUSAN WORTH JENKINS

Common city pigeons take refuge on a church's cross in Littleton. Even here they are scorned for their foul ways. Their nests are faintly visible behind the cross. The photographer's young daughter first spotted this dramatic scene.

PHOTOGRAPH LEGENDS

PAGE 140

A gull throws back its head and lets loose its shrill squeal amid the roar of traffic at I-25 and 6th Avenue. It is probably a ring-billed gull, the most common gull in (and on) Denver.

PAGE 141

A great horned owl is silhouetted on a lamppost in predawn light at the Denver Tech Center. Its night of hunting voles, mice, skunks and other small animals will soon give way to daytime slumber. Great horned owls are common around Denver. They are Colorado's earliest-nesting birds, starting to lay eggs in February.

PAGE 142

A ferruginous hawk keeps watch from an antenna at Rocky Mountain Arsenal. These largest and most regal of Colorado hawks are attracted to the arsenal especially by prairie dogs. Bald eagles, which usually don't hunt rodents, sometimes rob fresh kills from ferruginous hawks at the arsenal. This bird gets its name from its rusty shoulders, back and leggings.

PAGE 143

Turkey vultures migrating south in the fall relax at Rocky Mountain Arsenal. The abundance of wildlife at the arsenal should provide sufficient carrion for these scavengers. Turkey vultures are common summer residents in the southern part of the Denver area, such as at Castlewood Canyon and Roxborough State Parks. Their bare, red heads are distinctive.

PAGE 144 ABOVE

A sociable spider checks out condiments at the photographer's picnic.

PAGE 144 BELOW

Black-billed magpies are ubiquitous in the Denver area, even in places like Riverside Cemetery. Their abundance, noisy chatter and scavenging habits earn them indifference or disdain from some. Viewed objectively, however, these long-tailed, black and white birds with flashes of blue and green iridescence are strikingly beautiful.

PAGE 145

Only in Denver's Zoo would one find Dall sheep and pigeons together. Pigeons are highly adaptable, but they have not yet colonized the snowy, windswept crags of Alaska and British Columbia the sheep call home.

PAGE 146 ABOVE

A western meadowlark, perched on the simulated hogback of a Roxborough sign, flings its flutelike song skyward.

PAGE 146 BELOW

Pigeons perch on an ornate old bridge near downtown Denver's Confluence Park. Great flocks of these immigrants from Europe, a.k.a. rock doves, have adapted to life in Denver and cities everywhere. Their nests and droppings adorn the cityscape, inviting detractors.

PAGE 147

Two squirrels find warmth as the morning sun reflects off a metal roof at the Denver Tech Center.

PAGE 148

Bald eagles have little to learn about flying from the noisy aluminum bird taking off from Stapleton International Airport. This pair is perched at Rocky Mountain Arsenal with downtown Denver skyscrapers in the distance.

SHERM SPOELSTRA

This pronghorn seems reluctant to yield his territory, but once these houses are occupied by people, dogs and blaring TV sets, his kind will have to move further away from their traditional territory.

TIM CAZIER

TIM CAZIER

WENDY SHATTIL/BOB ROZINSKI

WENDY SHATTIL/BOB ROZINSKI

WENDY SHATTIL/BOB ROZINSKI

J. B. HAYES

J. B. HAYES

WENDY SHATTIL/BOB ROZINSKI

BILL BEVINGTON

WENDY SHATTIL/BOB ROZINSKI

WENDY SHATTIL/BOB ROZINSKI

WENDY SHATTIL/BOB ROZINSKI

ELEVEN

The Future
OF COLORADO URBAN WILDLIFE

JIM HEKKERS

The year is 2050. More than 85 percent of the people who live in Colorado are packed into the urban corridor that stretches the length of the Front Range from Fort Collins to Pueblo, spreads to Strasburg on the east, and reaches into the foothills and mountains to the west. From a distance in the air, this metropolis looks like any other in the west—Salt Lake City, Phoenix, Albuquerque, even—gasp—Los Angeles. A closer view, however, reveals differences. Bands of green surround the rivers and creeks winding through the metropolis. Large open expanses border the waterways and surround numerous ponds and lakes. The open areas in the eastern sections look very much like the prairie further east. Old-timers sometimes mutter about the relative lack of large areas of bluegrass.

For the tenth consecutive year, the Front Range metropolitan area is ranked as one of the three most livable such areas in the country. The ranking, as usual, has been helped considerably by high scores for environmental quality, a category that includes open space and wildlife. For a variety of reasons, the metropolitan area continues to have a range of urban wildlife species unmatched in the rest of the country. Moreover, conflicts between people and wildlife—particularly those that once involved deer, geese, prairie dogs, raccoons and skunks—are rare. The area also boasts a national distinction, the City and County of Denver's recent formation of a new city agency, the Denver Department of Urban Removal, representing the first effort by a major city to undo some development and restore it to open space.

Is this a feasible scenario for the future of urban wildlife in Colorado? Or is it merely overoptimistic drivel, wishful thinking spurred by hearing one too many environmental platitudes? Probably it's a little of both.

To get some idea about the future of urban wildlife in Colorado, specifically along the Front Range, I talked to a dozen local wildlife experts. Their views on the current situation and the direction we're heading varied considerably. Most, however, agreed on where we *ought* to be headed and, importantly, what we need to do in order to get there.

Where are we? The wildlife in and near our Front Range urban areas is in fairly good shape. As cities go, Denver and the rest of the Front Range cities are relatively young; we haven't had time to do too much irreversible damage to wildlife habitats. That's not to say we've been harmless. Our abuse of the South Platte River is tragic, as are narrow, concrete channels for other rivers built in the name of flood control. More often than not, our development schemes have ignored the needs of wildlife. And we seem linked with Kentucky bluegrass in a wasteful alliance.

Fortunately, our relative youth as a metropolis and our inherent abundance of wildlife have saved us from ourselves so far. We have even managed to take some very positive steps. The City of Boulder, Boulder County and Jefferson County, for example, are carrying out aggressive open space programs. Fort Collins is successfully integrating wildlife into its planning process. Many developers are recognizing the economic advantages of having open space and wildlife as part of their projects. As the need for and interest in water conservation grows, the stranglehold of bluegrass may be weakening. The state agency responsible for wildlife—the Division of Wildlife—is becoming increasingly aware of the needs and importance of urban wildlife, and cities and counties are beginning to listen to the agency's advice. Finally, and most importantly, a ground swell of public support for protecting our environment is becoming a political force.

Given the good news–bad news nature of the current situation, it's understandable that predictions about the future range from unbridled optimism to gloomy forecasts for a world of concrete, steel, bluegrass, smog and doom. It seems clear that the future of Colorado's urban wildlife will depend on how well we come to grips with a full range of complex and interrelated political, environmental, educational and economic issues.

The future of urban wildlife is linked to the future of our urban areas. In terms of sheer numbers of people, the prospects for our cities along the Front Range look like more of what we've already seen. Despite Colorado's current period of slower—and in some cases, stalled—growth, most population forecasters see a continuing upward trend. The majority of people will live in the urban areas along the Front Range. Already most Coloradans live in cities; some 55 percent live in the six-county Denver metro area, and nearly 82 percent of the state's 3.3 million residents live in urban areas. Colorado is expected to grow by more than one-third between now and 2010, and 90 percent of that growth is projected to be along the Front Range urban corridor.

Among people who care a lot about wildlife,

open spaces and the like, these numbers seem a bit frightening. How can wildlife possibly survive the onslaught of people and development? The answer is deceptively simple. Because some species—such as skunks and starlings—can adapt so easily, we will always have urban wildlife of some sort. The real issue is whether our wildlife will be an asset or a liability, something we enjoy or a nuisance. The answer to this question depends largely on how many people really do care about wildlife and what they do about it.

Judging by polls, my personal observations after dealing with people and urban wildlife issues for more than six years, and a host of other evidence, I think Colorado's citizens care deeply about wildlife. What's needed are two things: a way to translate that concern into actions and clear idea of what those actions should be.

My conversations with the experts addressed these issues. I've tried to condense their ideas into six action steps that could ensure that wildlife will be an asset in our urban environments 10, 50 and 100 years from now. These steps are:

- Preserve natural open spaces.
- Protect our drainageways and wetlands.
- Protect our water quality and quantity.
- Plant trees, shrubs and cover that provide food and shelter for wildlife.
- Improve our management of urban wildlife.
- Educate ourselves and our children about living in harmony with our environment.

All of these steps are important, but the most critical is the first—saving spaces for wildlife. That's because habitat is the key to the survival of wildlife. Some of the cities and counties along the Front Range are doing a commendable job of preserving natural open space; others lag far behind. Too often, planning does not include provisions for open space, or most of the so-called open space consists of bluegrass parks and sports fields. Cities need both parks and open space; the difference must be recognized. Bluegrass parks are good for soccer fields but lousy for wildlife.

Preserving open space requires a commitment at the basic planning level so that parcels are identified on the basis of wildlife needs rather than purely aesthetic and recreational needs. The planned community of Highlands Ranch in Douglas County is an excellent example of wildlife and open space being considered in the planning stage. Large open spaces in the community are being saved and left natural, as are corridors connecting the spaces. Preserving open space is relatively easy when development takes place on a large scale, but more often it occurs in small chunks. In these situations, local governments clearly need to identify open space needs and find the means to preserve the space, whether through outright purchase or through open space set-aside requirements for developers.

Competition for land in our urban areas is intense, so preserving space for wildlife can be expensive. Thus, funding mechanisms for the acquisition of open space are essential, especially in rapidly expanding areas such as Aurora. Sales taxes now fund open space programs in several areas. Other funding mechanisms—such as a check-off option on water bills—ought to be looked at too. We also ought to consider open space needs on a larger, regional scale. Better regional planning and coordination would ensure the right pieces in the right places. Further, we need to provide connections between open space parcels; usually these corridors are creeks, streams and rivers.

Riparian—river-bottom—land is Colorado's single most important type of habitat. It covers less than 3 percent of Colorado's land area, yet more than 90 percent of the wildlife species found in the state use it in some way. The story is no different in urban areas. Thus, how we handle the many creeks, streams and rivers that flow through the Front Range urban areas will be crucial to the future of urban wildlife.

These drainageways are often under pressure either directly from development or from the consequences of development—increased water runoff (because of the overlays of impermeable pavement) and the need to protect against flooding. Fortunately, a relatively recent trend in the treatment of waterways holds promise for the future. The philosophy that the only good drainageway is straight and made out of concrete is losing favor to a more natural approach. This approach usually requires a wider floodplain and more land than the classic concrete channel does, but its value to wildlife, and to people who appreciate trees and birds more than concrete, is immense.

Just as important to wildlife are the wetlands that surround not only rivers and streams but also ponds, lakes and low-lying areas. The value of wetlands and the consequences associated with their loss are being increasingly recognized. In fact, the commitment to protect wetlands extends from Washington, D.C., where President Bush has been vocal in his support, to local communities where citizen groups are closely watching developments and advocating for wetlands before planning commissions and city councils. Many cities are mapping their wetlands, an important first step in protecting them.

Wetlands provide many benefits, including their function as natural water filtering systems. This function is an important consideration, for the quality of our water is as critical to the survival of wildlife as it is to our own. Much of our wildlife lives under or on water, and many fish and other aquatic organisms are part of the food chains that support species that live on land.

Fortunately, because so many of our bodies of water are part of water supply and distribution systems, water quality is a high priority in

Colorado. The concern for quality needs to extend to lakes and ponds that are not part of the water supply system, such as flood control reservoirs. The formation of the Cherry Creek Basin Authority and the steps it has taken to lessen the pollution of Cherry Creek Reservoir by nutrient-rich and chemical-laden runoff from nearby housing and business developments illustrate the kind of attention water quality deserves in the future.

A more subtle but equally important water issue is that of quantity. Streams that are dry during part of the year or are reduced to trickles are less valuable to wildlife than streams that carry adequate water at all times. Assuring adequate minimum stream flows will be difficult, and the issues will be sorted out in complex arenas where water rights and water use questions are settled. The health of aquatic systems must be recognized as an important consideration when those questions are debated.

Since Colorado doesn't have an unlimited supply of water, finding ways to use it wisely will require a very difficult balancing act. We all will play a role in that process, especially in the area of water conservation. Will we accept the changes we need to make if we are to use less water? What about our bluegrass lawns? From a wildlife perspective, at least, less bluegrass would be welcome, especially if the grassy areas are replaced with vegetation more to the liking of birds and animals. Thus, the answer to the commonly asked question, "What can I do to make my yard more attractive to wildlife?" is simple: Reduce your bluegrass lawn and plant ground covers, shrubs and trees that provide food and shelter for the wildlife you want to attract. You'll certainly have a better chance of seeing more wildlife, and you're likely to use a lot less water.

Improving the quality of wildlife habitat in one or two yards per neighborhood will have very little effect on the overall health and quantity of urban wildlife. But if it's done on a large scale—an entire neighborhood, including parks and other public places, for instance—the benefits could be substantial. Plenty of resources are available to help you decide what to plant. The National Wildlife Federation, for instance, has an outstanding Backyard Habitat Program that deals with large-scale as well as small-scale projects.

The prospect of at least as much and maybe even more urban wildlife in the future is not without its disadvantages. Conflicts now between people and wildlife are common and sometimes unpleasant—skunks living under patios and porches, raccoons invading garbage cans, bats roosting in attics, geese grazing on golf courses, woodpeckers drumming on the siding of houses, starlings cackling by the thousands in backyard trees. These nuisances occur all too often to suit many people, mostly because of the lack of creative strategies to prevent problems or to deal with them once they happen. Solving such problems will require a greater emphasis on urban wildlife management.

Wildlife management has traditionally focused on game animals in wilderness and rural settings, largely because the source of wildlife management funding is hunting and fishing license fees. Also, the management activities that affect the most animals need to take place where the most animals are. However, in Colorado, as elsewhere, the interest in animals that don't fall into the game categories (collectively called nongame wildlife) has grown over the past 15 to 20 years. This change can be attributed in part to a gradual evolution in how wildlife in urban areas is viewed. Increasingly, urban wildlife management is being seen as an activity separate and distinct from managing wildlife in wilderness or rural settings.

In order for urban wildlife to remain a part of our lives, this evolution in attitude will need to continue and, if anything, accelerate. What's needed is more information about urban wildlife, as well as research into new management techniques. A good example is a project experimenting with live-trapping urban beavers and either moving them to remote locations or sterilizing them and putting them back where they were trapped.

Since they eat trees, beavers sometimes are at odds with people and agencies trying to improve waterways and parks by planting trees. In Denver, beavers have destroyed hundreds of thousands of dollars' worth of trees along Bear Creek and the South Platte River. Traditional wildlife management would simply prescribe trapping the beavers to keep their numbers under control and damage to a minimum. But most Front Range cities have passed laws against kill-trapping in response to both antitrapping sentiment and concerns about the safety of children and pets. In addition, the urban beaver habitat is so attractive that new beavers quickly move in to fill the territories of trapped beavers. The challenge to wildlife management is to find a way for beavers and trees to coexist.

The Denver beaver project will study the animals that are sterilized to see if they behave like normal beavers, maintaining the same family relationships and keeping other beavers out of their territories. The project may not find the solution, but it will at least provide more knowledge about urban beavers and move us closer to resolving the situation. Similarly, more research is needed into all of the species that tend to come into conflict with people and the things they build, grow or create.

How much progress is made in this area will depend on funding. Progress is relatively slow now because the state's wildlife management agency, the Division of Wildlife, is still primarily funded by hunting and fishing license revenues or by federal funds that come from excise taxes on hunting and fishing equipment. The only other funds received by the agency are contributions by taxpayers through the check-off box on state tax forms. This source raises anywhere from $300,000 to $400,000 or more a year, but the money must fund all nongame management activities. These include recovery efforts for threatened or endangered species, projects that carry a high priority and are expensive. Finding a source of funding for urban wildlife management will be a major challenge for the future.

Efforts to help citizens learn how to live in harmony with their environment pose an even greater challenge. The need for education covers a wide range of topics, from coexistence with wildlife to broad, complex issues such as global warming or the destruction of tropical rain forests. And the need is not simply for teaching at the school-age level; all of us must learn. Research indicates that people who live in urban areas tend to have less contact with and understanding of the natural world. No matter how citified we become, however, we will still be a part of the natural order of things, and it will be important that we understand and accept our role.

The Division of Wildlife and many urban school districts are responding to this challenge by bringing programs such as Project WILD into classrooms. Project WILD takes a multidisciplinary approach to teaching environmental concepts through innovative activities teachers can use in their classrooms. A number of large urban districts—Cherry Creek, Denver and Boulder, for example—have worked with the Division of Wildlife to integrate Project WILD into their curricula.

These types of educational efforts could be broadened and extended outside of school contexts. The Colorado Urban Wildlife Partnership, a coalition of public and private entities, is trying to focus attention on urban wildlife in order to stimulate interest across a broad spectrum and in many forums, from public libraries to parks and recreation districts to cultural facilities.

Without an emphasis on education, the rest of the steps I've outlined will be of little value. And without an interested, informed and motivated public the future not only of urban wildlife but of all wildlife and all natural landscapes like open spaces, wilderness, forests, marshes and the like is grim. The steps I've outlined are well within our power and means. If we are interested enough and willing to take action, we can have a future in which wildlife is an enjoyable aspect of city life. Not coincidentally, the steps we need to take to accomplish that end are also the steps we need to take to assure ourselves and our children a livable, high-quality future in our metropolitan areas.

ACKNOWLEDGMENTS

I wish to thank the following people for contributing information and expertise to this chapter: Steve Bissell, Ruth Carlson, Kathi Demarest, Beth Dillon, Lisa Evans, Phil Goebel, Mark Lamb, Rich Larson, Dave Lovell, Jeff Rucks, Pat Tucker and Dave Weber, Colorado Division of Wildlife; James C. Crain, City of Boulder Open Space; Craig Tufts, National Wildlife Federation; and Betsy Webb, Denver Museum of Natural History.

CATHERINE BARNES

If this black-tailed prairie dog had turned around, he would have seen his kind's future: eviction by commercial development along Denver's Arapahoe Road east of I-25.

TINA JONES

WENDY SHATTIL/BOB ROZINSKI

ABOVE: Birds seem willing to perch on almost anything that affords a good view. This black-crowned night-heron uses an old truck tire while fishing the South Platte River in Denver.
BELOW: A Canada goose scans its neighborhood from a high vantage point at Rocky Mountain Arsenal. In some places around Denver, elevated boxes and platforms are supplied for geese to use as nesting sites.

WENDY SHATTIL/BOB ROZINSKI

Because beaver are nocturnal, they are usually seen only through their work — gnawed and felled trees near ponds and streams. Besides chicken wire, an attempted solution to the "beaver problem" in the Denver area is to trap and sterilize the beasts. However, even neutered beaver need to eat and work.

Urban Wildlife PHOTO CLUB

Bini Abbott
Wayne Aigaki
Bob Ainsworth
Sandra Apel
Cecilia Armbrust
Barbara & Harold Arnold
Sonia & Jim Austin
Ray Austin
Bill Baker
William Bakke
Bob Barber
Catherine Barnes
Ron Beane
Bill Bevington
Ann Bonnell
Peter Botts
Pat & Lloyd Bowles
Dixie & Eddie Brewer
Dennis Brom
Helen Bronson
Karen Ann Bronson
Diane Buell
Terrell Campbell
Tim Cazier
Barbara Chase
Kent Choun
Susan Cody
Madonna & Cliff Cummings
Nadine & Bob Dean
Teresa DeDecker
Charles Delperdang
Donna & Aubrey De Woody
G. Richard Dickson, Jr.
Craig Dierksen
Denver Museum of Natural History
Zoology Department
Charles Duscha
Stephen Duvall
Marion & John Edwards
Jack Ferguson
Phyllis & Craig Fleming
Rose, Jan & Roland Fredrickson
George Frost
Joe Gallik
Judy & Dennis Gaskins
Elinor J. George
Scott Gilmore
Fred Gimeno
Irene Gimeno
Lorrie & Galen Gray
George Guest
Jim Haggard
Gary Haines
Judy Hall

Iris Hart
J. B. Hayes
Kelli Henrikson
Sannie Higdon
Linda Hoffman
Karen & John Hollingsworth
Richard Holmes
Paul B. Homan
Chuck Hursh
Ron Huston
Diane Hutton
Cathy & Gordon Illg
Susan Jenkins
Loraine Jones
Lynne, Dick & Jason Jones
Linda & Sam Jones
Tina Jones
Ira Kalfus
Deb Kanaga
Bob Kelso
Mary Kiesling
Carol Kirwin
Martin Kleinsorge
Ann & Dick Knutson
Fred Krampetz
Cecile & Chris La Forge
Lisa Lakel
Jack Laubach
Weldon & Scot Lee
Thomas Lentz
Fred Leonard
Mike Lockhart
Bonnie & Jerry Lofdahl
Joe Lusso
Gail McIntire
Del McNew
Beverly & Doug Mead
Meg Meyer
Vicki Milam
Dann Milne
Robert Mitchell
William Mosley
Bruce Nall
Jimmy Nelson
Sharon Nicholas
Phoebe & Doyal O'Dell
Roland Olderog
Betty & Harold Oliver
Jack Olson
Roger Owens
Esther & Lou Palmer
Catherine & Allan Passmore
Joanne Peachey
Sheryl & Dale Peterson
Steve Podgorski
Mr. & Mrs. Edward Potter
Dick Pratt
Chuck Purdy
Sandra Reay
Bruce Reinbold
Nancy & Kevin Reynolds
Lucille & Roger Riewerts
Laura Roark
Chuck Romeo
Bob Rozinski
David Sakaguchi
Robert Samland
D. Strat Saunders
Wendy Saunders
James Schultz
Betty Seacrest
Pat & Bill Searcy
Jean Settles
Linda & Calvin Shankster

Wendy Shattil
Evaline & Glen Shuster
Geraldine Slezak
Larry Smith
Robert Spencer
Sherm Spoelstra
Ron Standley
Donna & Joseph Stanley
John Stevenson
Terri Stewart
Sydney & Joel Storman
Jean Stout-Curlee
Kerry Strait
Joan Straits
Ellen Swanson
Shirley Swedeen
David Thele
Gary Thompson
Richard Thorpe
Virginia Trumpolt
Bob Usher
Else Van Erp
Rebecca & Paul Wacker
David Walker
Al Walls
Neil Ward
Betsy Webb
Mary Weight
Michael Weissmann
Emily & John Weller
Susan White
David Willette
Nancy Williamson
Kyla & Lloyd Wood

GARY D. HALL

PHOTOGRAPH LEGENDS

PAGE 164

Canada geese head for the Children's Zoo at Denver's Zoo.

PAGE 165 ABOVE

An American avocet wades at Walden Ponds near Boulder. Avocets feed by swishing their distinctively upturned bills through shallow water. They nest around some of the shallow ponds in the Denver area; this one raised two young in the summer of 1989.

PAGE 165 BELOW

It appears the Northern Migration Express dropped a load of male yellow-headed blackbirds near Barr Lake in the spring. Actually, some grain spilled here from a freight car, attracting the flashy birds. The males arrive from the south before the females to establish nesting territories.

PAGE 166 LEFT

A barn swallow poses prettily on a red stake in a golf course pond near the Denver Tech Center. From this perch, it will launch into an amazing aerobatic display as it catches flying insects.

PAGE 166 RIGHT

Could it be a Denver Broncos fan on the blue and orange picnic benches? A fox squirrel scans for leftovers at the picnic grounds in City Park just west of Denver's Zoo.

PAGE 167

A golden eagle warms to the rising sun at Rocky Mountain Arsenal. Notice the absence of wires on the pole that forms its perch. This pole is one of many erected at the arsenal just for raptors.

PAGE 168 ABOVE

Threat, insult, chatter, growl—what false bravado is afforded by a wire window screen! Face-offs like this must happen hundreds of times a day in Denver. Fox squirrel or Abyssinian kitten, which will turn tail first?

PAGE 168 BELOW

Six-year-old Alyson goes eye to eye with Nutkin, a typically bold, aggressive fox squirrel in east Denver. Such encounters have brightened many a morning for both.

PAGE 169 ABOVE

"Welcome to Burger King. May I take your order?" It took six weeks for the photographer to find these Canada geese in exactly the right spot once he visualized the shot.

PAGE 168 BELOW

Two denizens of Denver's Sloan Lake Park share a sunny park bench.

PAGE 170

An American robin nests on part of a dormant eighteen-wheeler. When the rig was needed, the nest had to be rescued. Even though nest moving directions from the Audubon Society and Colorado Division of Wildlife were followed, the bird abandoned the nest.

PAGE 171

An orb-weaving spider eyes its next meal, a billbug or weevil tangled in its web. This kind of spider likes to live under overhangs, such as awnings, patio decks, eaves or, as here, water faucets.

PAGE 172 LEFT

A fire hydrant in Aurora is a favorite perch for this burrowing owl; note the accumulation of droppings. The bird pauses here before taking food to its young. The nest is in a burrow in one of the many prairie dog colonies around Denver.

PAGE 172 RIGHT

Fear not! This cottontail washing its ear is in no danger of being flattened. It is only using a tractor parked for the winter at Rocky Mountain Arsenal as shelter from weather and predators. Rabbits are prey for raptors, coyotes, foxes and weasels; this one's chances of living more than a year are slim.

WENDY SHATTIL/BOB ROZINSKI

WENDY SHATTIL/BOB ROZINSKI

JOHN B. WELLER

WENDY SHATTIL/BOB ROZINSKI

WENDY SHATTIL/BOB ROZINSKI

WENDY SHATTIL/BOB ROZINSKI

WENDY SHATTIL/BOB ROZINSKI

JACK OLSON

WENDY SHATTIL/BOB ROZINSKI

DALE PETERSON

WENDY SHATTIL/BOB ROZINSKI

WAYNE AIGAKI

WAYNE AIGAKI

BILL BAKER

WENDY SHATTIL/BOB ROZINSKI

APPENDIX B

Checklist of Colorado's FRONT RANGE URBAN WILDLIFE

The following list contains the 385 common wildlife species of fish (54), amphibians (12), reptiles (21), birds (216) and mammals (82) that inhabit urban areas along Colorado's Front Range. Because they have not been inventoried to date, the list does not include the numerous invertebrates such as insects that are even more prevalent, nor does it include plants.

The fish species listed are found at elevations below 6000 feet. The amphibian, reptile, bird and mammal species are derived from the Colorado Division of Wildlife Latilong data. The Latilong system divides Colorado into 28 blocks, each defined by one degree in latitude and longitude. Each block measures roughly 50 by 70 miles, or 3500 square miles. Latilong blocks are numbered from left to right, and Colorado is divided into 4 rows and 7 blocks per row.

The species listed are those occurring in Latilong blocks 4, 11, 12 or 19, which have a seasonal abundance of "fairly common" or greater, and which occur in at least one of the following habitat types: ponderosa pine, scrub oak, mountain mahogany, short-grass prairie, mixed grasses–habitat alteration areas, aspen, riparian lowland, riparian transition, marshes, wet open ground, open water–streams, open water–lakes, agricultural areas, cropland and urban areas.

FISH

white sucker
western longnose sucker
hybrid grass carp
European carp
goldfish
stoneroller
northern redbelly dace
longnose dace
roundtail chub
creek chub
suckermouth minnow
fathead minnow
brassy minnow
plains minnow
common shiner
red shiner

sand shiner
bigmouth shiner
spottail shiner
golden shiner
mosquitofish
central plains killifish
plains topminnow
yellow perch
Johnny darter
Iowa darter
Arkansas darter
walleye
white bass
wiper
green sunfish
orangespotted sunfish
pumpkinseed
bluegill
hybrid sunfish
Sacramento perch
white crappie
black crappie
smallmouth bass
largemouth bass
brook stickleback
gizzard shad
rainbow smelt
northern pike
tiger muskie
Kokanee (sockeye) salmon
mountain whitefish
cutthroat trout
rainbow trout
brown trout
brook trout
channel catfish
black bullhead
brown bullhead

AMPHIBIANS

tiger salamander
plains spadefoot
New Mexico spadefoot
boreal toad
Great Plains toad
red-spotted toad
Woodhouse's toad
boreal chorus frog
plains leopard frog
bullfrog
northern leopard frog
wood frog

REPTILES

ornate box turtle
western painted turtle
western spiny softshell
eastern collared lizard
northern earless lizard
short-horned lizard
red-lipped prairie lizard
northern prairie lizard
northern many-lined skink
Great Plains skink
prairie-lined racerunner
Colorado checkered whiptail
eastern yellowbelly racer
western coachwhip
northern water snake
bullsnake
wandering garter snake

western plains garter snake
red-sided garter snake
northern lined snake
prairie rattlesnake

BIRDS

pied-billed grebe
horned grebe
eared grebe
western grebe
American white pelican
double-crested cormorant
great blue heron
snowy egret
black-crowned night-heron
white-faced ibis
snow goose
Canada goose
green-winged teal
mallard
northern pintail
blue-winged teal
cinnamon teal
northern shoveler
American wigeon
canvasback
redhead
ring-necked duck
lesser scaup
common goldeneye
bufflehead
common merganser
red-breasted merganser
ruddy duck
turkey vulture
Mississippi kite
bald eagle
northern harrier
sharp-shinned hawk
Swainson's hawk
red-tailed hawk
ferruginous hawk
rough-legged hawk
golden eagle
American kestrel
prairie falcon
ring-necked pheasant
blue grouse
wild turkey
northern bobwhite
scaled quail
Virginia rail
sora
American coot
sandhill crane
snowy plover
semipalmated plover
killdeer
mountain plover
American avocet
greater yellowlegs
lesser yellowlegs
solitary sandpiper
willet
spotted sandpiper
marbled godwit
semipalmated sandpiper
western sandpiper
least sandpiper
Baird's sandpiper
stilt sandpiper
long-billed dowitcher
common snipe

Wilson's phalarope
red-necked phalarope
Franklin's gull
Bonaparte's gull
ring-billed gull
California gull
herring gull
Forster's tern
black tern
rock dove
band-tailed pigeon
mourning dove
flammulated owl
eastern screech-owl
western screech-owl
great horned owl
burrowing owl
common nighthawk
common poorwill
chimney swift
black-chinned hummingbird
calliope hummingbird
broad-tailed hummingbird
rufous hummingbird
belted kingfisher
Lewis' woodpecker
red-headed woodpecker
Williamson's sapsucker
downy woodpecker
hairy woodpecker
olive-sided flycatcher
western wood-pewee
willow flycatcher
Hammond's flycatcher
dusky flycatcher
western flycatcher
Say's phoebe
ash-throated flycatcher
Cassin's kingbird
western kingbird
eastern kingbird
scissor-tailed flycatcher
horned lark
tree swallow
violet-green swallow
northern rough-winged swallow
bank swallow
cliff swallow
barn swallow
gray jay
Steller's jay
blue jay
scrub jay
pinyon jay
Clark's nutcracker
black-billed magpie
American crow
common raven
black-capped chickadee
mountain chickadee
plain titmouse
red-breasted nuthatch
white-breasted nuthatch
pygmy nuthatch
brown creeper
rock wren
Bewick's wren
house wren
American dipper
golden-crowned kinglet
ruby-crowned kinglet
blue-gray gnatcatcher
western bluebird
mountain bluebird

Townsend's solitaire
veery
Swainson's thrush
hermit thrush
American robin
gray catbird
northern mockingbird
sage thrasher
brown thrasher
water pipit
cedar waxwing
northern shrike
loggerhead shrike
European starling
Bell's vireo
solitary vireo
warbling vireo
orange-crowned warbler
Virginia's warbler
yellow warbler
yellow-rumped warbler
black-throated gray warbler
Townsend's warbler
Grace's warbler
American redstart
ovenbird
MacGillivray's warbler
common yellowthroat
Wilson's warbler
yellow-breasted chat
western tanager
black-headed grosbeak
blue grosbeak
lazuli bunting
green-tailed towhee
rufous-sided towhee
brown towhee
Cassin's sparrow
rufous-crowned sparrow
American tree sparrow
chipping sparrow
clay-colored sparrow
Brewer's sparrow
vesper sparrow
lark sparrow
black-throated sparrow
sage sparrow
lark bunting
Savannah sparrow
song sparrow
Lincoln's sparrow
white-crowned sparrow
Harris' sparrow
dark-eyed junco
McCown's longspur
Lapland longspur
chestnut-collared longspur
red-winged blackbird
western meadowlark
yellow-headed blackbird
Brewer's blackbird
common grackle
brown-headed cowbird
orchard oriole
northern oriole
rosy finch
pine grosbeak
Cassin's finch
house finch
red crossbill
pine siskin
lesser goldfinch
American goldfinch
evening grosbeak

MAMMALS

Virginia opossum
masked shrew
dusky shrew
water shrew
little brown myotis
Yuma myotis
long-legged myotis
western small-footed myotis
big brown bat
hoary bat
Townsend's big-eared bat
eastern cottontail
Nuttall's cottontail
desert cottontail
snowshoe hare
white-tailed jackrabbit
black-tailed jackrabbit
least chipmunk
Colorado chipmunk
yellow-bellied marmot
Wyoming ground squirrel
thirteen-lined ground squirrel
spotted ground squirrel
rock squirrel
golden-mantled ground squirrel
black-tailed prairie dog
Abert's squirrel
fox squirrel
red squirrel
Botta's pocket gopher
northern pocket gopher
plains pocket gopher
yellow-faced pocket gopher
silky pocket mouse
hispid pocket mouse
Ord's kangaroo rat
beaver
western harvest mouse
deer mouse
white-footed mouse
piñon mouse
rock mouse
northern grasshopper mouse
hispid cotton rat
eastern woodrat
Mexican woodrat
bushy-tailed woodrat
southern red-backed vole
meadow vole
montane vole
long-tailed vole
prairie vole
muskrat
Norway rat
house mouse
western jumping mouse
porcupine
coyote
red fox
swift fox
gray fox
black bear
ringtail
raccoon
marten
ermine
long-tailed weasel
mink
badger
western spotted skunk
eastern spotted skunk

striped skunk
mountain lion
bobcat
elk
mule deer
white-tailed deer
pronghorn
mountain sheep

ROLAND FREDRICKSON

The notorious bird-feeder bandit, a fox squirrel in Aurora goes to great length for bird seed. It is an immigrant to Colorado from the eastern United States.